An Introduction to Machine Learning in Quantitative Finance

Advanced Textbooks in Mathematics

Print ISSN: 2059-769X
Online ISSN: 2059-7703

The *Advanced Textbooks in Mathematics* explores important topics for postgraduate students in pure and applied mathematics. Subjects covered within this textbook series cover key fields which appear on MSc, MRes, PhD and other multidisciplinary postgraduate courses which involve mathematics.

Written by senior academics and lecturers recognised for their teaching skills, these textbooks offer a precise, introductory approach to advanced mathematical theories and concepts, including probability theory, statistics and computational methods.

Published

An Introduction to Machine Learning in Quantitative Finance
  by Hao Ni, Xin Dong, Jinsong Zheng and Guangxi Yu

Conformal Maps and Geometry
  by Dmitry Beliaev

Crowds in Equations: An Introduction to the Microscopic Modeling of Crowds
  by Bertrand Maury and Sylvain Faure

Mathematics of Planet Earth: A Primer
  by Jochen Bröcker, Ben Calderhead, Davoud Cheraghi, Colin Cotter, Darryl Holm, Tobias Kuna, Beatrice Pelloni, Ted Shepherd and Hilary Weller
  edited by Dan Crisan

Periods and Special Functions in Transcendence
  by Paula B Tretkoff

The Wigner Transform
  by Maurice de Gosson

Advanced Textbooks in Mathematics

An Introduction to Machine Learning in Quantitative Finance

Hao Ni
University College London, UK

Xin Dong
Citadel Securities LLC, UK

Jinsong Zheng
Huatai Securities, China

Guangxi Yu
SWS Research, China

World Scientific

NEW JERSEY · LONDON · SINGAPORE · BEIJING · SHANGHAI · HONG KONG · TAIPEI · CHENNAI · TOKYO

Published by

World Scientific Publishing Europe Ltd.

57 Shelton Street, Covent Garden, London WC2H 9HE

Head office: 5 Toh Tuck Link, Singapore 596224

USA office: 27 Warren Street, Suite 401-402, Hackensack, NJ 07601

Library of Congress Cataloging-in-Publication Data
Names: Ni, Hao (Lecturer in mathematics), author. | Dong, Xin, author. |
    Zheng, Jinsong, author. | Yu, Guangxi, author.
Title: An introduction to machine learning in quantitative finance /
    Hao Ni, University College London, UK, Xin Dong, Citadel Securities LLC, UK,
    Jinsong Zheng, Huatai Securities, China, Guangxi Yu, SWS Research, China.
Description: New Jersey : World Scientific, [2021] | Series: Advanced textbooks in mathematics,
    2059-769X | Includes bibliographical references and index.
Identifiers: LCCN 2020041545 (print) | LCCN 2020041546 (ebook) |
    ISBN 9781786349361 (hardcover) | ISBN 9781786349644 (paperback) |
    ISBN 9781786349378 (ebook) | ISBN 9781786349385 (ebook other)
Subjects: LCSH: Finance--Mathematical models. | Machine learning.
Classification: LCC HG106 .N52 2021 (print) | LCC HG106 (ebook) | DDC 332.0285/631--dc23
LC record available at https://lccn.loc.gov/2020041545
LC ebook record available at https://lccn.loc.gov/2020041546

British Library Cataloguing-in-Publication Data
A catalogue record for this book is available from the British Library.

For any available supplementary material, please visit
https://www.worldscientific.com/worldscibooks/10.1142/Q0275#t=suppl

Desk Editors: George Vasu/Michael Beale/Shi Ying Koe

Typeset by Stallion Press
Email: enquiries@stallionpress.com

This book is dedicated to my father Yannong Ni, who inspired me to write this book and is always supporting me.
Hao Ni

This book is dedicated to my family.
Xin Dong

This book is dedicated to my family.
Jinsong Zheng

This book is dedicated to my parents, Kongyi Yu and Lihua Wang, for their selfless support and encouragement.
Guangxi Yu

Preface

About Me

I am currently working in the financial mathematics group at University College London (UCL) with an affiliation of the Alan Turing Institute for data science and artificial intelligence. I work on interdisciplinary fields that include stochastic analysis, financial mathematics and machine learning. Most of my current research work is primarily in the area of sequential data mining, including financial data analysis, human–computer interface and computer vision.

I developed an interest in mathematics at a young age and earned my bachelor degree from Southeast University in China. In the last year of my undergraduate study, I joined the student exchange program to Ulm University, Germany and started to learn financial mathematics there. Then I read a master's degree in Mathematical and Computational Finance and my DPhil in mathematics from Oxford University, UK. During my doctoral time, I interned briefly at the quantitative teams at one insurance company in Munich and one investment bank in London. I continued to pursue my research interests as a postdoctoral researcher in the USA and the UK. I received two quantitative job offers upon the end of my postdoctoral work. But I chose to stay in academia, as I have the opportunity to explore a broad range of research areas with full freedom, and it seems more challenging and exciting to me. I began my faculty position at UCL in 2016, after four years of postdoctoral research at Brown University and the Oxford-Man Institute of Quantitative Finance.

During the post-doctoral period, my research interest gradually changed. Although my work was still related to theoretical mathematics, my main research area has shifted to machine learning and its applications. As you can see, I am not a typical machine learning researcher, who usually has a computer science background. Therefore, I always hope to have a chance to share my experience with more people who are new to machine learning and help them take fewer detours.

About This Book

I came up with the idea of organizing a series of events on machine learning for the first time in 2017, with the hope to help more people, especially those with a quantitative background, learn about machine learning. I have many friends and classmates working in the quantitative teams in financial industries, who are usually called "quants." Most of them have an educational background similar to me. They showed great interest in my current research, especially machine learning. However, due to busy work, they bear a considerable opportunity cost of self-study. Given that, I decided to organize a series of events to help them quickly understand the theoretical framework of machine learning, and to discuss current hot issues in financial mathematics and the potential applications of machine learning in finance.

In May 2018, with the help of friends, I organized six events as a first phase, including an introduction to machine learning, two lectures on supervised learning, two practice sessions and a financial case study.

Through these activities, I received a lot of valuable feedback and encouragement. All this urged me to go further and write a book based on the teaching materials for those events to help more readers who are interested in machine learning and quantitative finance to get a quick start. I hope that this book provides a general framework of machine learning and gives a rigorous but intuitive introduction to basic machine learning methods, with ample examples to enhance the reader's understanding. This book provides not only theoretical knowledge of machine learning, but also practical examples of financial applications. It aims to help readers have hands-on experience in this field and will hopefully enable them to apply what they learn to their own financial data problems.

The Python codes contained within this book have been made publicly available on the author's GitHub: `https://github.com/deepintomlf/mlfbook.git`.

About Machine Learning

There is no doubt that machine learning is a hot topic in both academia and industry. However, machine learning is not a panacea that you can use to solve problems by just feeding data into algorithms. Although the applications of machine learning have been successful in many areas, it is still far from human intelligence at present. This book aims to unveil the mysteries around machine learning. Most of the algorithms are built based on mathematical and statistical theories, which any senior undergraduate with a solid mathematical foundation can quickly master.

Financial mathematics has changed a lot in the last decade. Traditionally, financial mathematics is concerned with modeling the financial markets using stochastic models to solve pricing, hedging and trading problems in finance. Pricing and hedging using quantitative models are the main work of quantitative analysts in investment banks. As for those in funds, statistical methods have been taken as the principal method to find consistent trading signals and develop effective trading strategies. But in recent years, there has been more and more complex multimodal data available, which is potentially valuable in predicting future financial market movements. This trend has already motivated both the buy-side and the sell-side to invest a large amount of money in developing machine learning methods for mining such data. The Oxford-Man Institute of Quantitative Finance is part of the trend, and was jointly established by the Man Group and the University of Oxford, and became part of the machine learning group at the Department of Engineering at the University of Oxford in 2015.

In the future, the boundaries between disciplines will become vaguer and vaguer, and talents with interdisciplinary backgrounds will become increasingly needed. Hence, students who want to pursue a career in quantitative finance need to prepare for an interdisciplinary future and conform to the technical requirements of a changing world. They should have a solid mathematical background, expertise in machine learning and statistics, strong programming skills and domain knowledge in finance, as well as excellent communication skills.

I believe the development of machine learning or artificial intelligence has traditionally focused more on the engineering side, which has led to huge successes in various domains, such as image recognition and robotics. Currently, many machine learning researchers and data scientists mainly focus on winning state-of-the-art competitions or improving the performance of models on certain datasets or applications. There is nothing wrong with that, since machine learning is an applied discipline. For the improvement of numerical results in the short term, tuning parameters may be more effective than establishing mathematical foundations for machine learning algorithms. However, in the long run, it is crucial to understand why those methods were so successful and how to improve them further. I believe mathematics can offer an answer to this question, and provide a solid theoretical foundation to data science as a whole. On the one hand, a good understanding of algorithms is helpful to tuning parameters and modifying models efficiently. On the other hand, working on real applications helps theoretical researchers to abstract important mathematical questions that have a huge impact in real life and to think more about the complexity and improvement of algorithms from a more systematic perspective. Theories and applications are complementary to each other, and should not be a divide.

About the Future

Curiosity and the ability for persistent learning are the most important factors for an individual's personal development. I have an educational background in mathematics. During my PhD study of three years, I always thought that I only enjoyed mathematics because of its elegance, beauty and complexity. In the meantime, I thought that programming and applications were easy. Nevertheless, as I have gained more and more interdisciplinary research experience, I have realized how ignorant I was before. So do not easily conclude what you like and dislike about things you do not actually know. Most of the time, it is the fear of the unknown rather than aversion. For fields with which we are not familiar, we should stay open minded and curious to learn more about them, which may bring potential opportunities to us in the future.

To conclude, I would like to quote from Russell—one of my favorite sentences:

> *Three passions, simple but overwhelmingly strong, have governed my life: the longing for love, the search for knowledge, and unbearable pity for the suffering of mankind.*

Hao Ni

About the Authors

 **Hao Ni** is an associate professor of financial mathematics at University College London (UCL) and a Turing Fellow at the Alan Turing Institute since September 2016. Prior to this, she was a visiting postdoctoral researcher at ICERM and Department of Applied Mathematics at Brown University from September 2012 to May 2013. She continued her postdoctoral research at the Oxford-Man Institute of Quantitative Finance until 2016. She finished her DPhil in mathematics at the University of Oxford. Her research interests include stochastic analysis, financial mathematics, and machine learning. More specifically, she is interested in non-parametric modeling effects of complex multi-modal data streams through rough paths theory and machine learning. Moreover, she has research interests in real-world applications, such as human–computer interface, computer vision and finance.

 **Xin Dong** is currently a quantitative researcher working at Citadel Securities in London since 2018. She was a desk strategist at Morgan Stanley in London from 2014 to 2018. Prior to that, she finished her PhD focused on point processes in the Department of Mathematics at Imperial College London in 2014. Her current research interests are in quantitative trading.

Jinsong Zheng is an algorithm engineer at Huatai Securities in China since 2019. Prior to this, he was a quantitative risk manager at Talanx AG in Germany. He got his DPhil in economics at the University of Duisburg Essen. He has extensive work experience in quantitative risk management and financial modeling. His current research interests are quantitative investment and asset allocation, especially using machine learning techniques and big data.

Guangxi Yu is a quantitative analyst at SWS Research in China. His current research includes financial data analysis, quantitative modeling and equity strategies. He is especially interested in machine learning applications in finance. Prior to this, he obtained his Master's degree in Financial Mathematics at University College London under the supervision of Dr. Hao Ni.

Acknowledgments

First of all, I offer my deepest gratitude to my co-authors of the book—Xin Dong, Jinsong Zheng and Guangxi Yu. I initially wrote the central part of the book in English based on the slides of my meetup events on machine learning in quantitative finance. Xin added two chapters on unsupervised learning, while Jinsong contributed to the implementations of RRL and the code samples for the case study. Guangxi translated the whole book to the Chinese version and helped with editing and proofreading the English version of the book. My co-authors are all full-time employees in quantitative finance and have busy work lives. In the past five months, they devoted most of their spare time to working with me on this book project. Without their contribution, this book would not exist in its current form.

Besides this, I want to especially thank my friends Xin Dong, Lan Jiang, Cuiyu He, Jing Liu and Hanyi Chen. I organized a series of activities on machine learning. All these friends patiently corrected my speech drafts for each event, which were the embryo of this book.

Furthermore, I am also grateful to my friends who helped me with the event venue for these activities. I want to thank Charlotte Zhang from Wind Information and Wen Xing from the Singapore Exchange for providing a venue for the first activity. Thanks to Xing Zhang, a participant of the first activity from Citigroup Inc., for offering an excellent place for later events at Citigroup Inc.

I have a long list of people who helped with the proofreading and provided valuable suggestions—Patricia Andrew, Jun Guo, Cuiyu He, Richard Hoyle, Siran Li, Jing Liu, Jiazi Tang, Mingjing Wu, Weixin Yang and others. I owe a debt of gratitude to all of them.

I would also thank my students Shujian Liao and Hang Lou for helping me with some of the numerical results in Chapters 5 and 9.

Huge thanks to the World Scientific Publishing, in particular Laurent Chaminade, Michael Beale, George V and Zahra Asgarali.

Moreover, I would like to acknowledge that my research is funded by the EPSRC under program grant EP/S026347/1 and by the Alan Turing Institute under EPSRC grant EP/N510129/1.

Finally, I want to thank my loving family, who are always supporting me to follow my heart and pursue my dreams. I would like to pay my special thanks to my father, as he encouraged me to write a book based on the presentation slides of my meetup events and contacted Tsinghua University Press for the publication of my book. Without him, I would not have been motivated to finish this book in both Chinese and English versions so quickly.

Disclaimer

The contents of this book solely reflects the analysis and views of the authors. No recipient should interpret this document to represent the general views of Citadel Securities. Facts, analyses, and views presented herein have not been reviewed by and may not reflect information known to other Citadel Securities professionals.

Contents

Listings

Chapter 1

Overview of Machine Learning and Financial Applications

1.1 Big Data Era

The world is undergoing a data explosion. We are indeed in the era of big data. Looking around at all kinds of data, one may wonder and ask: what is big data? In contrast to ordinary data, big data refers to data with the 3-v characteristics, i.e., volume, velocity and variety (Figure 1.1).

(1) **Volume** defines the size of data. In the past few years, the total amount of data has increased enormously. Both business processes and individual activities have been creating datasets of large volume. For example, modern telescopes produce terabytes of data per observation and simulation, which is the data size required to model our observable universe and can easily push supercomputers to their limits.

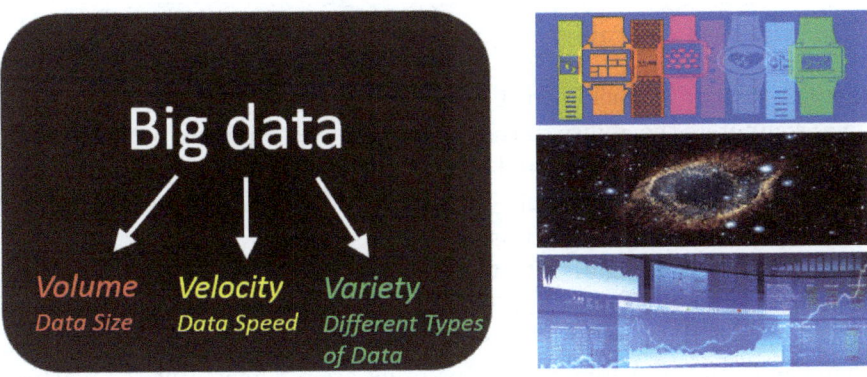

Figure 1.1. 3-v characteristics of big data.

(2) **Velocity** defines the speed of data streams. Nowadays, data streams can arrive at an unprecedented speed and must be dealt with in a timely manner. High-frequency trading is an excellent example of that, and a high-frequency financial data stream can come every millisecond.

(3) **Variety** means various kinds of data, which come in different types of formats from structured, numeric data in traditional databases to unstructured text documents, emails, videos, audios, etc. In the sub-figure on the top right corner of Figure 1.1, you see different types of wearables. Smartphones are must-haves for individuals. Those devices record our data of various kinds, including photos, text messages, online shopping transactions, etc.

The 3-v characteristics have posed many challenges that conventional statistical methods cannot handle. But meanwhile, it provides many potential opportunities to make better decisions using big data, as recent advancements in both hardware and software enable us to develop technologies to extract useful information from big data. Machine learning is an ideal candidate to address the challenges and exploit the opportunities.

Machine learning can efficiently process large datasets and extract useful information. Let us give you one example of a standard dataset in machine learning research so that you have a sense of the data scale in machine learning applications. An advanced machine learning algorithm can successfully recognize an online handwritten character at an accuracy of over 95% using the CASIA-OLHWDB1 dataset, which has 4,037 categories (3,866 Chinese characters and 171 symbols) and 1,694,741 samples, produced by 420 writers.

1.2   Machine Learning

In 1959, Arthur Samuel defined machine learning as "a field of study that gives computers the ability to learn without being explicitly programmed." Sometimes it is difficult to fulfil some kinds of task by giving explicit programming instructions. In this case, the best approach is to enable computers to have the ability to learn from the data.

As we can see in Figure 1.2, image recognition is an example of a machine learning task. In this task, it is difficult to tell the computer how to recognize flowers in paintings by explicit programming instructions as the position, shape and color of flowers in each painting are different. But the machine learning algorithm can learn which flower is in the paintings automatically if

Labels: **Iris flower**   **Poppy flower**   **Sun flower**

Figure 1.2. Image classification in machine learning.

Machine Learning is a field of study to give computer systems the ability to "learn" with data, without being explicitly programmed.

*Machine learning is ideal for exploiting the opportunities hidden in **big data**.*

Figure 1.3. The motivation of machine learning.

we feed the algorithm with enough data of paintings and the corresponding flower labels. The same algorithm can be applied to identify a human pose in photographs. Here you see that machine learning algorithms can be applied to different tasks and do not need to be explicitly programmed for each specific task.

Machine learning algorithms are data-driven and operated by building a model from observations of data to predict an unseen situation without much human intervention. They are well suited to the complexity of dealing with disparate data sources. Unlike traditional analysis, machine learning thrives on growing datasets. The more data is fed into a machine learning system, the more it can learn and the better results it can get (Figure 1.3). Therefore machine learning is ideal for addressing the challenges and exploiting the opportunities in big data.

A core objective of a machine learning algorithm is to generalize from its experience. In this context, generalization is the ability of a learning machine to perform accurately on unseen data based on the observation of a learning dataset. It is important to note that a sophisticated machine learning algorithm can get high accuracy in the training set. However, this does not guarantee high predictive accuracy on unseen data due to the potential risk of the *overfitting* issue. The overfitting phenomenon is usually caused by the misinterpretation of noise as signal by over-sophisticated algorithms, which leads to poor predictive power. Therefore, to build appropriate machine learning algorithms, we need to take into account both the model complexity and data size.

Machine learning algorithms can be mainly divided into supervised learning, unsupervised learning and reinforcement learning.

- **Supervised Learning** aims to learn from a labelled training set such that the resulting model can be effectively applied to unseen data. Starting from the analysis of a given training dataset, which is a collection of input–output pairs, the supervised learning algorithm produces an inferred function to make predictions about the output values.
- **Unsupervised Learning** studies how systems can infer a function to describe a hidden structure from unlabeled data.
- **Reinforcement Learning** (RL) is a learning method to identify optimal actions that interacts with its environment by maximizing the corresponding rewards. This method allows machines and software agents to automatically determine the ideal behavior within a specific context based on a feedback loop from the environment.

Table 1.1 provides a summary of the main categories of machine learning and corresponding examples of financial applications.

In contrast to supervised learning, unsupervised learning algorithms are used when the data is neither classified nor labeled. As compared to supervised learning and unsupervised learning, RL is different in terms of goals. While the goal in unsupervised learning is to find similarities and differences between data points, the goal in reinforcement learning is to find a suitable action model that would maximize the reward of the agent. However,

Table 1.1. Categories of machine learning and financial application examples.

- Supervised Learning
  - Regression (e.g., forecast returns)
  - Classification (e.g., predict the direction of returns)
- Unsupervised Learning
  - Clustering (e.g., identify the most common signs of market stress)
- Reinforcement Learning (e.g., learning trading strategy)

supervised learning aims to learn the mapping between the input and labeled output from the data.

Let us go back to the trading example, in which you may see the difference between RL and supervised learning. Uncovering the pattern between market information and future price movement (see Section 5.4.7) is a supervised learning problem. In contrast, the optimal trading problem is about finding the best trading strategy (action) to maximize final wealth (reward), which falls into the reinforcement learning category.

The recent successful applications of machine learning have been mainly attributed to the advancement of hardware technology, which has made the storage, processing and analysis of big data by machine learning algorithms feasible in practice. The successes in machine learning applications are mainly focused on supervised learning and reinforcement learning. Here we enumerate several examples of machine learning applications. One of the most popular machine learning algorithms, so-called deep learning, has achieved state-of-the-art results in various tasks, such as image recognition and speech recognition.

- Supervised Learning: image recognition and speech recognition [LeCun *et al.* (2015a)].
- Reinforcement Learning: Atari video game [Mnih *et al.* (2013)], AlphaGo [Silver *et al.* (2016)].

1.3 Quantitative Finance

1.3.1 Challenges of financial data

With the advent of the big data era, we have seen a dramatic increase in the range of available real-time financial market datasets, from online transaction records to high-frequency limit order books (a record of outstanding buy or sell orders). Extracting information from these financial datastreams using machine learning techniques is challenging due to their low signal-to-noise ratio and complex multi-modality. Those issues may cause the misinterpretation of noise in the data as signals, and thus lead to potential financial loss, and even financial crises.[1]

---

[1] At the Alan Turing Institute, I lead a research project on analyzing noisy data stream with focused application in finance. Interested readers refer to https://www.turing.ac.uk/research/research-projects/analysing-noisy-data-streams for more information on this project.

Besides, an increasing amount of unstructured data—e.g., the financial news, satellite images for earth observation data or investment forum chat— is potentially valuable to provide useful information to financial services. However, traditional statistical analysis cannot process and analyze these unstructured data, which are also called alternative data.

Furthermore, some financial data may be limited in terms of data availability—e.g., some financial instruments exist only for a short period, which may lead to insufficient data for sophisticated machine learning techniques. Last but not least, financial data may not be stationary, which means that they may undergo regime changes and render older data less relevant for prediction [Sirignano and Cont (2018)].

1.3.2   Recent development of machine learning for financial applications

In recent years, more and more financial companies have started to adopt advanced machine learning techniques to gain an edge in market competition. Increasingly many hedge funds have moved away from traditional analysis methods and have adopted machine learning algorithms for predicting fund trends and portfolio selection.[2] An article entitled Seven Ways Fintechs Use Machine Learning to Outsmart the Competition explain various successful applications of machine learning in the Fintech sector to improve financial analysis.[3]

A perfect example of this is JPMorgan Chase's COIN, which stands for contract intelligence. It takes about 360,000 hours for humans to review 12,000 commercial credit agreements, while the COIN analyzes legal documents and contracts using image recognition software, completing its task in a matter of seconds.[4]

As the quantity of data available and the access to it have grown, leveraging machine learning to make better investment or other business decisions has been validated in various applications in finance. This increases people's faith in the great potential of advanced machine learning in finance. Machine learning provides an efficient tool to analyze data, ranging from

---

[2]https://www.jpmorgan.com/insights/research/machine-learning.

[3]https://igniteoutsourcing.com/fintech/machine-learning-in-finance/.

[4]https://www.independent.co.uk/news/business/news/jp-morgan-software-lawyers-coin-contract-intelligence-parsing-financial-deals-seconds-legal-working-a7603256.html.

Table 1.2. Machine learning solutions to address financial data challenges.

Financial Data Challenges	Machine Learning Solutions
Noisy data	Model ensemble, data augmentation, filtering, cross validation...
Heterogeneity	Change-point detection...
New data type	Can handle unstructured data, e.g., text data and image data.

individual data (e.g., sentiment from news, Twitter) to the business processes (e.g., credit history). Table 1.2 summarizes the machine learning solutions to key financial data challenges we have discussed.

1.3.3 The future of quantitative finance

In the future, the trend of adoption of machine learning in quantitative finance is inevitable. *Interpretability, data fusion and hardware technology* are the most dominant trends of machine learning for the sustainable advancement of quantitative finance.

Interpretability of black box learning algorithms is an emerging research area. It aims to enable people without a technical background to have more confidence using black-box algorithms and to check the results produced by the algorithms using their domain knowledge. Zhang and Zhu (2018) provides a review of recent studies in understanding the representations of neural networks, which are one of the most popular algorithms and are discussed in Chapter 5. Although (deep) neural networks have exhibited superior performance in terms of prediction accuracy in various tasks, interpretability may be their weakest point. Gunning (2017) provides several slides to summarize the recent development of explainable machine learning, including visualization and interpretability of machine learning algorithms. The explainability of machine learning algorithms is very important in financial applications. Building confidence in machine learning algorithms for investors and regulators is a must for widespread utilization of machine learning in finance. Explainable machine learning plays a critical role in building such confidence and trust.

The second important angle of future machine learning research in quantitative finance is to design efficient algorithms for data fusion. Finance related data can be of various types, e.g., news data, transaction history, etc. How to extract useful information from diverse types of financial data is critical to investigate.

Lastly, machine learning cannot advance further without the development of hardware and high-performance computing. Most of the deep learning algorithms were proposed decades ago, but their successful applications have mainly emerged in the last few years. This is because, at the time of introducing those algorithms, computing technology had barely been ready to realize the power of deep learning algorithms. Looking forward, data sizes are growing exponentially and algorithms are becoming more and more sophisticated, which requires large scale computing technology to meet the needs. Distributed systems, quantum computing, etc. are promising research areas with huge potential and impact.

1.4   Next Generation of Talents in Quantitative Finance

An article entitled What should we be teaching the next generation of quants? KNect365's Finance[5] provides a nice description of the skills of future talents in quantitative finance, who are called "quants." See the full article at the below link.[6]

The first essential skill is a solid understanding of mathematics and statistics, which is a must for quants. Traditionally, quants should have a strong background in stochastic calculus, probability theory and financial mathematics. As such, a doctoral degree in some quantitative fields such as mathematics, statistics and physics are very desirable by most firms. In the last decade, however, graduate students in computer science and machine learning have gained increasing popularity. One would expect that the expertise of machine learning will be highly desirable in the future.

Second, quants should command adequate financial market knowledge. One may think that machine learning or artificial intelligence may downplay human knowledge on a specific domain. But the reality is that those algorithms are far from smart. Therefore domain knowledge is essential for quantitative finance, as it will help to identify the "right" and critical

---

[5]https://finance.knect365.com/.

[6]https://knect365.com/quantminds/article/9efba3f6-6271-4584-acac-4747d28235 c5/what-should-we-be-teaching-the-next-generation-of-quants.

questions to ask and provide the insights to guide the search for quantitative solutions.

Third, strong programming skills are also critical for quants to perform at a proficient level. C++ is one of the most popular programming languages and is commonly used for high-frequency trading. Recently, with the rise of machine learning, Python is getting more and more popular as it provides numerous libraries for statistical learning and data analysis that are easier to learn and use than those in C++.

The last, but not least, skill is communication. An ideal candidate for quant has not only strong analytic expertise but also excellent interpersonal skills. The future talents should have the ability to communicate with people from different backgrounds and to link business needs with sophisticated machine learning algorithms. With the development of machine learning, more and more repetitive work will be automated and eventually replaced by algorithms—e.g., algorithmic trading. However, decision-makers in the financial markets are still humans, and the success of real-world applications of machine learning relies on how effectively the quants communicate the results provided by the algorithms to their colleagues in the front offices, clients or financial regulators.

1.5 Outline of the Book

We summarize the outline of this book in Table 1.3. First of all, we start with an overview of the development of machine learning and its

Table 1.3. Structure of the book.

(1) Overview of machine learning and quantitative finance
(2) Supervised learning:

- General framework
- Linear regression and regularized linear regression
- Non-linear regression (basis expansion)
- Tree-based methods: decision tree, random forest and gradient boosting trees
- Neural network: ANN, CNN and RNN

(3) Unsupervised learning: cluster analysis and PCA
(4) Reinforcement learning
(5) Case study in finance

applications in finance in Chapter 1. Then in Chapter 2, we introduce a general framework of supervised learning. In Chapter 3, we describe the simplest linear regression and the regularization method, e.g., Lasso and Ridge Regression. Following this, we turn to discuss non-linear regression/classification methods. In Chapter 4, we study tree-based models for supervised learning. We devote Chapter 5 to a discussion of neural networks, which focuses on the three main kinds of neural network, i.e., Artificial Neural Network (ANN), Convolutional Neural Network (CNN) and Recurrent Neural Network (RNN).

We introduce unsupervised learning in Chapters 6 and 7, including cluster analysis and principal component analysis (PCA). In Chapter 8, we focus on reinforcement learning and its application in portfolio optimization. Chapter 9 is a case study in finance and discusses one popular data challenge in credit score prediction, using the machine learning methods covered in the previous chapters to equip the readers with hands-on experience with practical data problems.

In each chapter, we introduce the mathematics of the different learning algorithms and aim to deliver the core ideas behind each algorithm concisely and rigorously. Then numerical examples in Python code are provided after each theory section to help readers better understand the methodology and develop hands-on machine learning expertise to be able to solve real-world problems. The key contents of each session are summarized in a table with a bounded box, to help readers review each session. All supplemental material (e.g., code examples) is available for download at `https://github.com/deepintomlf/mlfbook.git`.

1.6   Useful Resources

There is a vast amount of educational material and software packages on machine learning available. Here we introduce some useful Python libraries for machine learning, recommended books or reading materials, and online platforms for data competitions or working demo collections.

1.6.1   Python libraries

In this subsection, we briefly introduce several useful Python libraries that are extensively used in machine learning.

- NumPy
- Pandas

- Matplotlib
- Scikit-Learn
- Keras

NumPy[7] is the fundamental package for scientific computing with Python, which is useful to provide a powerful multi-dimensional array object and advanced tools for integrating C/C++ and Fortran code. It has useful functionalities to handle linear algebra, Fourier transforms and random number generation.

Pandas[8] is an open-source library providing high-performance, easy-to-use data structures and data analysis tools, which enable users to implement data structures and operations for numerical tables and time series.

Matplotlib[9] is a library for plotting quality figures in various formats and interactive environments across platforms. For example, one can use it to create a histogram or box-plot of data.

Scikit-Learn[10] is an open-source toolbox built on Numpy, Scipy and matplotlib, which provides simple and efficient tools for data analysis and contains most of popular machine learning algorithms.

Keras[11] is a high-level neural network API, capable of running on top of TensorFlow, CNTK, or Theano. It aims to enable fast experimentation in deep learning and can run with both CPU and GPU. One main strength of Keras is to allow users to achieve fast prototyping. It is very easy for beginners to start their first deep learning project (e.g., recognizing the pen-digits in images using MNIST data) in a few minutes using Keras.

1.6.2 Books and other online reading materials

(1) Methodology and Theory of Machine Learning:

- Friedman, J., Hastie, T., and Tibshirani, R., 2001. *The Elements of Statistical Learning.* New York: Springer series in statistics.
- Goodfellow, I., Bengio, Y., Courville, A., and Bengio, Y., 2016. *Deep Learning* (Vol. 1). Cambridge: MIT press.
- Rasmussen, C.E. and Williams, C.K., 2006. *Gaussian Process for Machine Learning.* MIT press.

---

[7]https://www.numpy.org/.
[8]https://pandas.pydata.org/.
[9]https://matplotlib.org/.
[10]https://scikit-learn.org/stable/.
[11]https://keras.io/.

(2) Hands-on Machine Learning with Python Programming:

- Géron A., 2017. *Hands-on Machine Learning with Scikit-Learn and TensorFlow: Concepts, Tools, and Techniques to Build Intelligent Systems.* "O'Reilly Media, Inc." [12]

If one aims to develop hands-on machine learning skills, the only way is by doing machine learning applications oneself. Kaggle and AI Hub are two good online platforms for data science and related case studies. Kaggle[13] is an online community of data scientists and machine learners, which is owned by Google LLC. It allows users to find and share datasets, explore and build models in a web-based data-science environment, and enter competitions with the collaborators to solve data challenges. Besides, it provides an excellent public data platform and a cloud-based workbench for data science, which are particularly useful for beginners to learn machine learning. The case study in finance covered in Chapter 8 is a recent Kaggle competition on Home Credit Default Risk Prediction. Once selecting a Kaggle competition to enter, one may learn many useful things from the public kernel (python codes) shared by other participants, including common tricks on data cleaning, feature engineering and model ensemble.

AI Hub[14] provides a collection of Interactive Machine Learning Examples. For example, a python notebook entitled Explaining the predictions of a black-box model with Shapley[15] works on energy price predication as a concrete example and explains how to use SHAP (Shapley Additive Explanations) to interpret results for a black-box model that is hosted on the Google Cloud Machine Learning Engine (CMLE). AI Hub provides interactive machine learning demos using Google Colaboratory,[16] which is a free Jupyter notebook environment that requires no setup and runs entirely in the cloud. For collaborative projects, Colaboratory enables you to write and execute code, save and share documents, and access powerful computing resources without the hassle of setting up the correct version of Python.

---

[12]Example code and solutions to the above: https://github.com/ageron/handson-ml.
[13]https://www.kaggle.com/.
[14]https://aihub.cloud.google.com/.
[15]https://aihub.cloud.google.com/p/products%2F18634c01-4dac-48e7-b104-a0e727cb979b.
[16]https://colab.research.google.com/.

1.7   Go Beyond This Book

This book provides an introduction to various machine learning methods, but there are still many mainstream machine learning algorithms that we do not discuss here. For example, we do not present any probabilistic methods (Bayesian type algorithms) due to the limited length of the book. However, Bayesian inference is very important for quantifying model uncertainty, and is worthy of study.

In addition to the standard machine learning methods discussed in this book, there are several advanced methods. We list a few prevalent and advanced machine learning methods as follows, but this list is not exhaustive.

- Generative Adversarial Networks (GANs) were firstly proposed in [Goodfellow *et al.* (2014)]. GANs have enormous potential, as they can learn to mimic any distribution of data in theory. They have successfully been used to generate fake data, impressively similar to empirical image data and speech data. For example, WaveNet[17] is a generative model for raw audio [Van Den Oord *et al.* (2016)].
- Transfer learning aims to solve a machine learning problem about how to apply knowledge gained while solving one task to a different but related task.
- Manifold learning provides non-linear dimension reduction based on the assumption that the high dimensional data is embedded in the lower dimensional manifold—locally data lives in a low dimensional Euclidean space [Ache and Warren (2019)].
- Meta-learning (learn to learn [Thrun and Pratt (2012)]) aims to generalize to new tasks that have never been encountered during training time. Based on previous experience on a variety of learning tasks, it learns how to complete the new tasks.

We sincerely hope that this book serves to equip readers with a solid theoretical basis for studying advanced machine learning methods effectively, and to apply them to solve real-world data problems. This is just the beginning of your adventure in the data world.

---

[17]https://deepmind.com/blog/wavenet-generative-model-raw-audio/.

Chapter 2

Supervised Learning

Supervised learning is the machine learning task of inferring a function that maps an input to an output based on example input–output pairs. Depending on whether the output is a continuous variable or a categorical variable, the supervised learning can be further divided into two types:

- Regression;
- Classification.

In Section 2.1, we first focus on regression problems. A general framework of regression includes the model, loss function, optimization, prediction and validation. Each component of the regression framework is discussed in detail in the following sections. In Section 2.2, we explain how to go from regression to classification. Lastly, we discuss how to use an ensemble of multiple models to enhance the performance of supervised learning.

2.1 Framework of Regression

Let us introduce the standard setup of the regression problem. For concreteness, we consider the case of a scalar output. Suppose that we have the dataset $\mathcal{D} = \{(x_i, y_i)\}_{i=1}^N$, where (x_i, y_i) denotes the ith input–output pair (also called the ith sample). Each sample input x_i is a d-dimensional vector, i.e., $x_i := (x_i^{(1)}, \ldots, x_i^{(d)}) \in \mathbb{R}^d$. Assume that there exists $f : \mathbb{R}^d \to \mathbb{R}$, such that

$$y_i = f(x_i) + \varepsilon_i, \tag{2.1}$$

where $y_i \in \mathbb{R}$ and ε_i are independent and identically distributed (iid) random variables with $\mathbb{E}[\varepsilon_i | x_i] = 0$. For the regularity assumption of f, assume

15

that f is a continuous function. For ease of notation, we also adopt the matrix form for $\mathcal{D} = (X, Y)$, where

$$X = \begin{pmatrix} x_1^{(1)}, x_1^{(2)}, \ldots, x_1^{(d)} \\ \vdots \quad \vdots \qquad \vdots \\ x_N^{(1)}, x_N^{(2)}, \ldots, x_N^{(d)} \end{pmatrix} \text{ and } Y = \begin{pmatrix} y_1 \\ \vdots \\ y_N \end{pmatrix}, \tag{2.2}$$

where X is an $N \times d$ matrix and Y is an $N \times 1$ vector.

The first question we ask is how to estimate the corresponding output for any given new input x_*. In the context of the regression problem, a rigorous mathematical formulation of this question is to estimate $\mathbb{E}[y|x = x^*]$, i.e., $f(x^*)$ for any given new input data x^*, which is equivalent to estimating f. Thus f is also called the *mean* function of the regression problem.

The next important question is how to choose the best estimator for f among different possible estimators, which boils down to what "best" means and how to quantify the performance of each estimator.

In the following, we explain how to approach the above two questions and summarize this as a general framework for regression. Recall that the goal of the regression problem is to learn the fixed but unknown mean function f from the labeled dataset $\mathcal{D}$ such that Equation (2.1) holds. A natural step is to postulate the model f_θ to describe the unknown mean function f, where θ are the model parameters that fully characterize the model f_θ. In this way, the problem of finding f is translated into finding the best parameters θ to fit the data.

To find the best parameters θ, we need to quantify what we mean by "the best parameters." Motivated by this, we propose the *loss function* to quantify the discrepancy between the model estimated output $f_\theta(x)$ and the actual output y. Once choosing the loss function $L(\theta|\mathcal{D})$, the optimal parameter set θ^* is defined to be the one that minimizes the loss function. In most cases, there is no closed formula for the optimal parameter set θ^*, and we need to use the numerical optimization method. No matter how we obtain the estimator of the optimal parameters θ^*, either by closed formula or numerical methods, once we have θ^*, we are ready to make prediction. More specifically, for any new input x_*, the estimator of the conditional expectation of the output $\mathbb{E}[y_*|x_*]$ is given by $f_{\theta^*}(x_*)$. Lastly, we need to quantify the goodness of the fit by specifying the metrics, e.g., mean squared error (MSE), R-squared (R^2). Those metrics may not be the same as the one used in the loss function.

Table 2.1. The framework of regression.

Dataset:	$\mathcal{D} = \{(x_i, y_i)\}_{i=1}^{N}$
Model:	$f_\theta(x) \approx \mathbb{E}[y|x] = f(x), \; \forall x \in \mathbb{R}^d$
Empirical Loss:	$L(\theta|\mathcal{D}) = \frac{1}{N}\sum_{i=1}^{N} d(f_\theta(x_i), y_i) \rightarrow$ Minimize
Optimization:	$\theta^* = \arg\min_\theta(L(\theta|\mathcal{D}))$
Prediction:	$\hat{y}_* = f_{\theta^*}(x_*)$
Validation:	Compute the indicators for the goodness of fit

Table 2.1 summarizes the entire process that we described above. Dataset, model, empirical loss, optimization, prediction and validation are the key elements of supervised learning. We follow this general framework to introduce several supervised learning algorithms in the following chapters and summarize each algorithm in the framework box.

In the rest of the chapter, we discuss each component of the framework, including model, loss function, optimization and prediction/prediction in details.

2.1.1 Model

In this subsection, we introduce various types of models, ranging from linear models to non-linear models and explain the main idea behind most non-linear models—so-called basis expansion. In regression, the proposed model is a family of parametric functions, say f_θ, where θ denotes the parameter set, which fully characterizes the model. For simplicity, we focus on the one-dimensional output case.

Let us start with the simplest model—the linear model—where we assume that $f_\theta \colon \mathbb{R}^d \to \mathbb{R}$ is a linear function, i.e., $\forall x = (x^{(1)}, x^{(2)}, \ldots, x^{(d)}) \in \mathbb{R}^d$,

$$f_\theta(x) = \theta^T x = \sum_{j=1}^{d} \theta^{(j)} x^{(j)},$$

where $\theta = (\theta^{(1)}, \ldots, \theta^{(d)}) \in \mathbb{R}^d$ is the parameter set. This is the model adopted by linear regression methods (Chapter 3).

However, linear models might not be rich enough to describe the complex functional relationship between the input and the output. Motivated by this, there are various types of non-linear models. We list some popular non-linear models as follows, but the list is not exhaustive.

◇ Polynomial model, e.g.,

$$f_\theta(x) = x\mu + x\Sigma x^T,$$

where $\theta = (\mu, \Sigma)$, and $\mu \in \mathbb{R}^d$ and $\Sigma \in \mathbb{R}^d \times \mathbb{R}^d$.

◇ Spline model, e.g.,

$$f_\theta(x) = \sum_{i=1}^{M} C_i(x - l_i)^+,$$

where $\theta = (l_i, C_i)_{i=1}^{M}$ are model parameters.

◇ Regression tree model (Chapter 4):

$$f(x) = \sum_{m=1}^{M} c_m \mathbb{I}(x \in R_m),$$

where $\{R_1, R_2, \ldots, R_M\}$ is a partition of the input space with M disjoint regions. The tree model allows the partition of the input space by splitting variables and points, which agrees with the topology that a tree should have (e.g., Figure 2.1).

◇ Neural network model (Chapter 5).

Neural network models are based on a collection of connected neurons (nodes). There are various types, which are illustrated in Figure 2.2. We elaborate the main types of neural network models in Chapter 5.

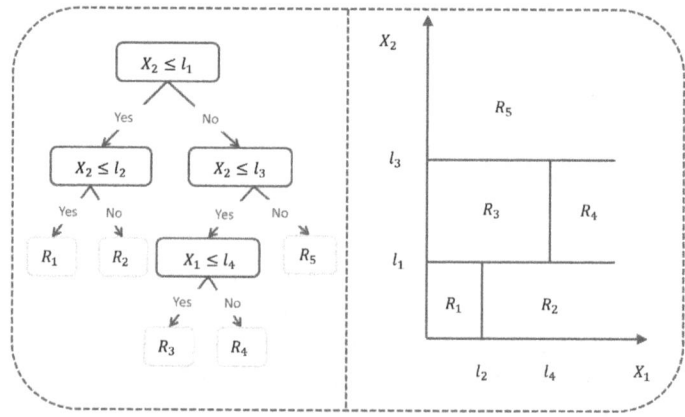

Figure 2.1. An example of a tree model.

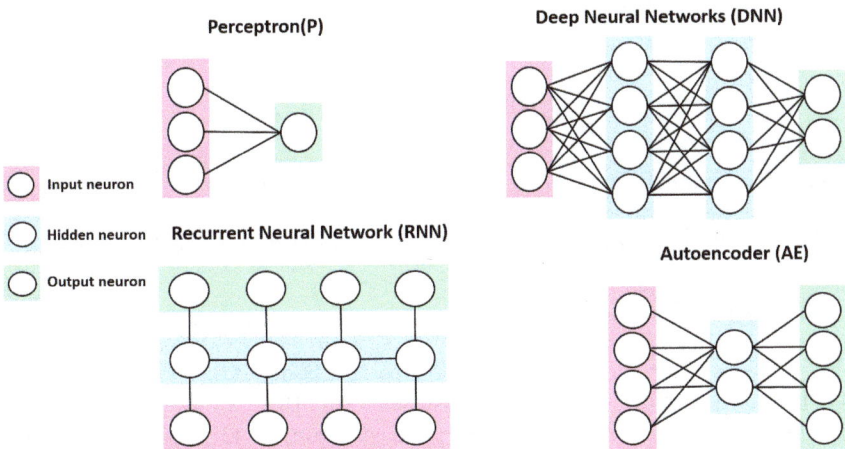

Figure 2.2. Examples of the main types of neural network architectures.

2.1.2 Loss function

In statistics, the loss function (also called cost function) is proposed to quantify the difference between estimated and actual values for output data. It serves as a utility function for parameter estimation. The loss function is a measure for parameters. The smaller the value of the loss function, which indicates that the estimated output is closer to its actual output, the better the parameter is.

The concept of the loss function represents the price paid for inaccuracy of predictions in learning problems. One of the most commonly used loss functions in regression is the quadratic loss function, which is defined as the squared error between the model estimated output and the actual output (see Definition 2.1).

Definition 2.1 (Quadratic Loss Function). Let f_θ denote the model fully characterized by parameters θ. The quadratic loss function is defined to be that $\forall (x, y) \in E \times \mathbb{R}$,

$$Q_\theta(x, y) = (y - f_\theta(x))^2.$$

We can evaluate the loss function for each sample. Averaging the loss function of all samples leads to the empirical risk, which denotes the average loss on the whole data set.

Definition 2.2 (Empirical Risk). Let Q_θ denote a loss function where θ is a model parameter set. Then the empirical risk denoted by L is defined as follows:

$$L(\theta|\mathcal{D}) = \frac{1}{N} \sum_{i=1}^{N} Q_\theta(x_i, y_i),$$

where $\mathcal{D} = (x_i, y_i)_{i=1}^{N}$.

In the following we often call the empirical risk the loss function.

2.1.3   Optimization

After specifying the loss function $L(\theta|\mathcal{D})$, the next step is to find the optimal parameter set θ^* to minimize the loss function. In general, unlike for standard linear regression (Ordinary Least Squares, OLS for short), there is no closed formula for the optimal parameters $\hat{\theta}$. It is important to design an effective numerical algorithm to find the optimal parameters. There are various numerical methods for optimization methods, including

- gradient descent based methods;
- gradient boosting method;
- expectation–maximization method.

In this section, we focus on the gradient descent based methods. The gradient boosting method is left for discussion in Chapter 4 and the expectation–maximization method (EM) is covered in Chapter 6.

Gradient descent is a first-order iterative optimization algorithm for finding the minimum of a function, which can be applied to tackle numerical optimization. We start with the gradient descent (GD) method and explain the main idea and intuition behind it. Batch gradient descent (BGD) is an algorithm for employing GD to estimate the optimal parameters to minimize the loss function. Then we discuss the variants of GD to accommodate the computational issues caused by large scale datasets by introducing randomness to the GD, i.e., stochastic gradient descent (SGD) and mini-batch gradient descent (mini-batch GD). Those methods are particularly widely used for the neural network models discussed in Chapter 5.

2.1.3.1   *Gradient descent method*

Gradient descent (GD) is a general first-order iterative algorithm to solve optimization problem numerically, which can find the local optimal $\hat{\theta}$ such that $\hat{\theta}$ achieves the local minimum of a given differentiable function $f :$ $\mathbb{R}^p \to \mathbb{R}$. The main idea is to find a local minimum of a function using GD by taking steps that are proportional to the negative of the gradient of the function at the current point.

Intuitively, imagine that you are lost in the mountains in a dense fog, and you only feel the slope of the ground below your feet. A reasonable strategy to get to the bottom of the valley quickly is to go downhill in the direction of the steepest slope. Mathematically what we aim to do is to construct a convergent sequence of $(\theta_n)_{n=0}^{\infty}$ such that

$$\theta^* = \lim_{n\to\infty} \theta_n, \tag{2.3}$$

where θ^* is a local minimum. A sufficient condition for such $(\theta_n)_{n=0}^{\infty}$ is that

(a) there exists an integer N_0 large enough such that $(f(\theta_n))_{n\geq N_0}$ is a non-increasing sequence w.r.t. n.

(b) when $\lim_{n\to\infty} \theta_n = \theta^*$,

$$\lim_{n\to\infty} \nabla f(\theta_n) = 0, \tag{2.4}$$

where $\nabla f(\theta)$ is the derivative of f at θ, i.e., $\nabla L(\theta) = (\partial_{\theta_1} L(\theta), \cdots, \partial_{\theta_p} L(\theta))$.

(c) the derivative of f is continuous.

When $\nabla L(\theta)$ is continuous, then condition (b) implies that

$$\nabla f(\theta^*) = \nabla f(\lim_{n\to\infty} \theta_n) = 0, \tag{2.5}$$

i.e., $f(\theta^*)$ is the local minimum of f.

In the GD algorithm, at the $(n+1)^{th}$ iteration, for given θ_n, we update the $(n+1)^{th}$ estimator θ_{n+1} by

$$\theta_{n+1} = \theta_n - \eta \nabla L(\theta_n),$$

where $\eta > 0$ is a constant, which is also called the learning rate and will be discussed in detail below. Next, let us explain why the above update can fulfill the sufficiency condition.

(a) When η is small enough, by Taylor's expansion,

$$L(\theta_{n+1}) - L(\theta_n) \approx \nabla L(\theta_n) \underbrace{(\theta_{n+1} - \theta_n)}_{-\eta\nabla L(\theta_n)} = -\eta \left(\nabla L(\theta_n)\right)^2 \leq 0.$$

It follows that for some integer $N_0 > 0$, $(L(\theta_n))_{n\geq N_0}$ is a non-increasing sequence as the above equation holds when the first order Taylor expansion holds.

(b) Suppose that $\{\theta_n\}$ is a convergent series, then

$$\lim_{n\to\infty} \theta_{n+1} = \lim_{n\to\infty} \theta_n - \eta \lim_{n\to\infty} \nabla L(\theta_n)$$

$$\Downarrow \qquad\qquad \Downarrow$$

$$\theta^* = \theta^* - \eta \lim_{n\to\infty} \nabla L(\theta_n).$$

It follows that $\lim_{n\to\infty} \nabla L(\theta_n) = 0$.

The GD algorithm is often called the *steepest gradient descent*. Let us explain to you the reason behind this name. By Taylor expansion, we have that

$$L(\theta) \approx L(\theta_0) + \nabla L(\theta_0)(\theta - \theta_0).$$

In the above Taylor expansion approximation, $L(\theta)$ decreases fastest on the optimal direction, which is equivalent to the minimization of $\nabla L(\theta_0)(\theta - \theta_0)$. We can show that the gradient direction $\nabla L(\theta_0)$ is the optimal direction, given the constraint that the distance between θ_0 and θ is a positive constant η. Mathematically, it is equivalent to show that if θ^* is the solution to the following constraint optimization problem,

$$\hat{L}(\theta) := \nabla L(\theta_0)(\theta - \theta_0) \to \min, \tag{2.6}$$

$$\text{subject to } ||\theta - \theta_0||_2 = \eta, \tag{2.7}$$

then there exists $\lambda_* \in \mathbb{R}$ such that

$$\theta^* = \theta_0 - \lambda_* \nabla L(\theta_0),$$

where $\lambda_* = \frac{\eta}{||\nabla L(\theta_0)||_2}$.

Proof. This constraint optimization problem can be rewritten as an unconstrained problem using the Lagrange multiplier:

$$\tilde{L}(\theta, \lambda) = \nabla L(\theta_0)(\theta - \theta_0) - \lambda(||\theta - \theta_0||_2^2 - \eta^2) \to \min, \tag{2.8}$$

where $\lambda \in \mathbb{R}$.

Then the optimal (θ^*, λ^*) satisfies that

$$\nabla \tilde{L}(\theta^*, \lambda^*) = 0.$$

Thus we have that

$$\nabla L(\theta_0) - 2\lambda^*(\theta^* - \theta_0) = 0. \tag{2.9}$$

By rearranging Equation 2.9, we have the formula for θ^* as follows:

$$\theta^* = \theta_0 + \frac{1}{2\lambda_*} \nabla L(\theta_0).$$

It is noted that as λ^* is a scalar, the optimal direction θ^* from θ_0 is along the gradient of $\nabla L(\theta_0)$.

Table 2.2. Summary of the gradient descent (GD) method.

Goal:	Find the local optimum θ^* to minimize a continuously differentiable function L.
Algorithm:	Initialize θ_0. For $n = 1 : N_e$, $\theta_{n+1} = \theta_n - \eta \nabla L(\theta_n)$, where N_e is the maximum number of iterations and η is the learning rate.
Idea:	We construct a sequence of $\{\theta_n\}_{n \geq 0}$ such that • For some N, $(L(\theta_n))_{n \geq N}$ is a decreasing sequence, i.e., $\quad L(\theta_N) \geq L(\theta_{N+1}) \geq \cdots$; • $\lim_n \nabla L(\theta_n) = 0$. This implies that $\{\theta_n\}_{n \geq 0}$ converges to the local minimum θ^*.

The only remaining part is to find the scalar λ^*. Equation 2.7 ensures that

$$||\theta^* - \theta_0||_2 = \frac{1}{2|\lambda^*|} ||\nabla L(\theta_0)||_2 = \eta. \qquad (2.10)$$

Thus it implies that $2|\lambda|^* = \frac{1}{\eta} ||\nabla L(\theta_0)||_2$. Then we have that $\lambda^* = \pm \frac{1}{2\eta} ||\nabla L(\theta_0)||_2$. Thus there are only two possibilities for λ^*, which is either $\frac{\eta}{2} \nabla L(\theta_0)$ or $-\frac{\eta}{2} \nabla L(\theta_0)$. It follows that

$$\hat{L}(\theta^*) = \begin{cases} \eta, & \text{if } \lambda^* = \frac{1}{2\eta} \nabla L(\theta_0); \\ -\eta, & \text{if } \lambda^* = -\frac{1}{2\eta} \nabla L(\theta_0). \end{cases} \qquad (2.11)$$

Recall that the goal is to find θ^* that minimizes $\hat{L}(\theta)$. Thus it implies that $\lambda^* = -\frac{1}{2\eta} \nabla L(\theta_0)$ and

$$\lambda_* = -\frac{1}{2\lambda^*} = \frac{\eta}{||\nabla L(\theta_0)||_2}. \qquad \Box$$

The summary of GD is given in Table 2.2.

2.1.3.2 *Discussion on learning rate*

The learning rate η is an important hyperparameter in the GD algorithm. A hyperparameter is a model parameter whose value is set before the learning process begins. By contrast, the parameters of the model can be trained from data, like θ. Most machine learning algorithms require hyperparameters.

Figures 2.3a and 2.3b show that there is a trade-off in the scale of the learning rate: when the learning rate is too small, the convergence of the parameters $(\theta_n)_n$ might be relatively slow; however, if the learning rate is

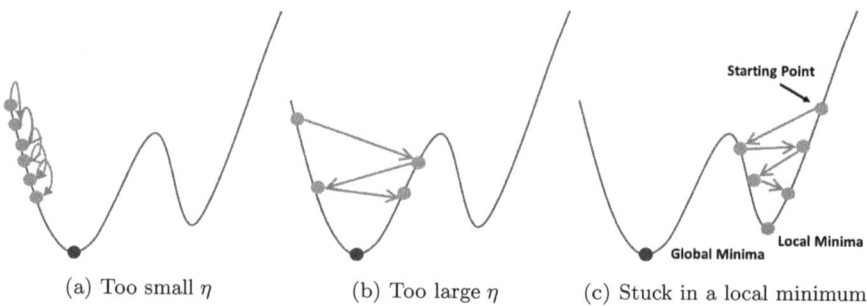

(a) Too small η (b) Too large η (c) Stuck in a local minimum

Figure 2.3. Effects of learning rate η.

too large, there may be the possibility that $(\theta_n)_n$ is bouncing between two valleys, which may also take a very long time to converge.

It is important to note that the GD algorithm cannot ensure a global minimum in a general setting, which makes the initialization of the parameters and learning rate important. The GD algorithm may be stuck at some local minimum, which is depicted in Figure 2.3c. In this case, a sufficiently large learning rate can help with escaping the local minimum.

2.1.3.3 *Batch gradient descent*

Batch gradient descent (BGD) is an algorithm for applying GD to minimize the empirical loss function; the update rule of BGD requires the computation of the gradient of the empirical loss function evaluated for *all the examples in the training set*. Let us recall the empirical loss function $L(\theta|\mathcal{D})$, which is usually in the additive form,

$$L(\theta|\mathcal{D}) = \frac{1}{N} \sum_{i=1}^{N} Q_\theta(x_i, y_i).$$

Thus the gradient of $L(\theta|\mathcal{D})$ with respect to θ is simply

$$\nabla_\theta L(\theta|\mathcal{D}) = \frac{1}{N} \sum_{i=1}^{N} \nabla_\theta Q_\theta(x_i, y_i).$$

A summary of BGD is given in Table 2.3.

Note that the gradient term is computed across all the samples in the dataset. One cycle through an entire training dataset is called an epoch. Therefore, it is often said that BGD performs model updates at the end of

Table 2.3. Summary of BGD.

Goal:	Find optimal θ such as to minimize $L(\theta\|\mathcal{D})$ in the form: $L(\theta\|\mathcal{D}) = \frac{1}{N}\sum_{i=1}^{N} Q_\theta(x_i, y_i)$.
Algorithm:	Initialize θ_0. For $n = 1 : N_e$, $\theta_{n+1} = \theta_n - \eta \underbrace{\nabla L(\theta_n\|\mathcal{D})}_{\text{Gradient term}}$ $= \theta_n - \eta \underbrace{\frac{1}{N}\sum_{i=1}^{N} \nabla_\theta Q_{\theta_n}(x_i, y_i)}_{\text{Gradient term}}.$
Idea:	Direct application of GD to empirical loss function.

Table 2.4. Pros and cons of BGD.

Pros

- Stable convergence: BGD may require reduced model update frequency because it has a more stable error gradient at each iteration.
- The computation of the gradient term can be implemented in a parallel manner.

Cons

- A too stable error gradient may result in convergence of the model to a local minimum, which is a less optimal set of parameters.
- At the end of the training epoch the updates require the additional complexity of accumulating prediction errors across all training examples.
- BGD is usually implemented in such a way that the entire training dataset is stored in memory and available to the algorithm. Thus BGD is memory-greedy and has very slow model updates for large datasets.

each training epoch. The advantages and disadvantages of BGD are summarized in Table 2.4.[1]

2.1.3.4 *Stochastic gradient descent*

Stochastic gradient descent, or SGD for short, is a GD-based algorithm that calculates the error and updates the model for *each example* in the training dataset. The main difference between BGD and SGD is the update rule for

---

[1]https://machinelearningmastery.com/gentle-introduction-mini-batch-gradient
-descent-configure-batch-size/.

each iteration: In SGD, for each iteration, the update of the model is based on the derivative of $L(\theta|\mathcal{D})$ w.r.t. θ evaluated at a randomly chosen sample in the training set—i.e., at each step $n \geq 1$, given θ_n, we update θ_{n+1} by

$$\theta_{n+1} = \theta_n - \eta_n \underbrace{\nabla_\theta Q_{\theta_n}(x_{i_n}, y_{i_n})}_{\text{Stochastic gradient term}},$$

where the index i_n is randomly selected from $\{1, \ldots, N\}$. The update of the model has the randomness of choosing the training example, which explains 'stochastic' in the name of SGD. SGD is also often called an online machine learning algorithm. Next let us explain the intuition behind SGD without worrying about the technical difficulty of proving its validity. Recall the empirical loss $L(\theta|\mathcal{D})$, which satisfies that

$$L(\theta|\mathcal{D}) = \frac{1}{N} \sum_{i=1}^{N} Q_\theta(x_i, y_i). \tag{2.12}$$

SGD randomly chooses a sample from the dataset to calculate the gradient. Suppose that i_n are iid with uniform distribution, where $n \in \{1, 2, \ldots, N\}$. Then it follows that

$$\mathbb{E}_{x,y}[\nabla_\theta Q_\theta(x_{i_n}, y_{i_n})] = \nabla L(\theta|\mathcal{D}), \tag{2.13}$$

where (x_{i_n}, y_{i_n}) is sampled from the empirical distribution of $(x_i, y_i)_{i=1}^N$. Or alternatively we can sample (i_n) randomly from $\{1, \ldots, N\}$ without replacement. Equation (2.13) still holds. It implies that although for each iteration the stochastic gradient term is not $\nabla L(\theta|\mathcal{D})$, its expectation coincides with $\nabla L(\theta|\mathcal{D})$. As the number of the maximum iteration N_e tends to infinity, it is reasonable to expect that the limit of θ_n by SGD converges to the local minimum as that of GBD, which shares the spirit of Monte Carlo methods.

However, to make the SGD algorithm work, we need to adjust the learning rate by choosing a suitable decreasing step-size sequence $\{\eta_n\}_n$ instead of the constant learning rate η in BGD. The reason for this is that the limit of the stochastic gradient term cannot converge to zero if there is a gradient evaluated for at least one sample that is non-zero. However, to ensure the convergence of θ_n, we have to make the sequence of η_n converge to zero. This explains why in SGD, the learning rate η_n needs to be reduced gradually. A summary of SGD is given in Table 2.5.

Table 2.5. Summary of SGD.

Goal:	Find optimal θ such as to minimize $L(\theta|\mathcal{D})$ in the form: $L(\theta|\mathcal{D}) = \frac{1}{N}\sum_{i=1}^{N} Q_\theta(x_i, y_i)$.
Algorithm:	Initialize θ_0. For $n = 1 : N_e$, Randomly choose the index i_n from $\{1, \cdots, N\}$, $\theta_{n+1} = \theta_n - \eta_n \underbrace{\nabla_\theta Q_{\theta_n}(x_{i_n}, y_{i_n})}_{\text{Stochastic gradient term}}$, for a suitably chosen decreasing step-size sequence $\{\eta_n\}_n$.
Idea:	$\mathbb{E}_{x,y}[\nabla_\theta Q_\theta(x_{i_n}, y_{i_n})] = \nabla L(\theta|\mathcal{D})$, where (x_{i_n}, y_{i_n}) is sampled from the empirical distribution of $(x_i, y_i)_{i=1}^{N}$ or randomly sampled without replacement.

Table 2.6. Pros and cons of SGD.

Pros

- The increased model update frequency may result in faster learning on some problems.
- Noisy gradient updates can avoid the premature convergence of the model to local minima.

Cons

- Updating the model so frequently is more computationally expensive than other configurations of gradient descent. SGD may take significantly longer to train models on large datasets.
- The frequent updates can result in a noisy gradient signal, which may cause the model parameter updates have a higher variance over training epochs and in turn make the model error more oscillatory.
- The unstable estimate of the error gradient can also make it difficult for the algorithm to settle on an error minimum for the model.

Let us summarize the benefits and downsides of SGD in Table 2.6.[2]

To sum up, SGD is very quick to evaluate each iteration. Randomness helps to escape a local minimum, but it makes the settling of the minimum difficult.

---

[2]https://machinelearningmastery.com/gentle-introduction-mini-batch-gradient-descent-configure-batch-size/.

2.1.3.5 *Mini-batch gradient descent*

Mini-batch gradient descent (mini-batch GD) is another variant of the gradient descent algorithm, which splits the training dataset into small batches that are used to calculate model error and update model coefficients. Mini-batch GD can be viewed as a combination of BGD and SGD.

At one iteration, instead of going over all examples, mini-batch GD updates the gradients based on a subset of samples (called mini-batches) for the given batch size b. When $b = 1$, mini-batch GD is SGD; when $b = N$, mini-batch GD is BGD. In mini-batch GD, the typical method of creating mini-batches includes two steps:

(1) Shuffle the dataset to avoid the existing order of samples.
(2) Split the entire training data set into several non-overlapping mini-batches of batch size b; if the sample size is not divisible by the batch size, the remaining samples will be their own batch.

Then we apply batch gradient descent for each mini-batch until all the samples have been processed (this is called one epoch); we repeat this procedure until the number of epochs reaches the maximum epoch number N_e.

Implementations may take average of the gradient, which further reduces the variance of the gradient. Mini-batch gradient descent aims to strike a balance between the robustness of stochastic gradient descent and the efficiency of batch gradient descent. In the field of deep learning (Chapter 5), mini-batch GD is the most common optimization method used to estimate the optimal model parameters.

A summary of mini-batch GD is given in Table 2.7.

Table 2.7. Summary of mini-batch GD.

Goal:	Find optimal θ such as to minimize $L(\theta\|\mathcal{D})$ in the form: $L(\theta\|\mathcal{D}) = \frac{1}{N}\sum_{i=1}^{N} Q_\theta(x_i, y_i)$.

Algorithm:	Initialize θ_0.
	For $n = 1 : N_e$,
	    Randomly partition the dataset $\mathcal{D}$ into $N_b = \frac{N}{b}$ mini-batches of size b, denoted by $(B_i)_{i=1}^{N_b}$.
	    For $j = 1 : N_b$, $$\theta_{n+1} = \theta_n - \eta_n \underbrace{\frac{1}{b} \sum_{(x,y) \in B_j} \nabla_\theta Q_{\theta_n}(x, y)}_{\text{Stochastic gradient term}},$$ where $\{\eta_n\}_n$ is a suitably chosen decreasing sequence.

Idea:	Combining SGD and BGD.

Table 2.8. Pros and cons of mini-batch GD.

Pros
• The model update frequency is higher than batch gradient descent, which allows for a more robust convergence and avoids local minima.   • The mini-batch updates provide a computationally more efficient process than SGD.   • The mini-batch algorithm allows a balance of both the efficiency of not having all training data in memory and algorithm implementations.
Cons
• Mini-batch requires an additional "mini-batch size" hyperparameter for the learning algorithm. It may increase the computation cost as this hyper-parameter needs to be tuned in practice.   • Error information must be accumulated across mini-batches of training examples, as for batch gradient descent.

The advantages and disadvantages of mini-batch GD are listed in Table 2.8.[3]

2.1.3.6 *Comparison of three types of gradient descent*

In the previous subsections, we have discussed three gradient descent methods, i.e.,

- Batch GD;
- Stochastic GD;
- Mini-batch GD.

They vary in terms of the number of training samples used to calculate empirical loss and to update the model. A summary of the comparison between the above three methods is provided in Table 2.9. We can see that there is a trade-off between the computational efficiency of gradient descent configurations and the accuracy of gradient updates. Figure 2.4 depicts the typical trajectory of parameter sequence for these three methods. More optimization methods can be found at the following website.[4]

---

[3]https://machinelearningmastery.com/gentle-introduction-mini-batch-gradient-descent-configure-batch-size/.

[4]http://ruder.io/optimizing-gradient-descent/.

Table 2.9. Comparison of various GD based methods.

	BGD	SGD	Mini-batch GD
Update frequency	Low	High	Medium
Update complexity	High	Low	Medium
Fidelity of error gradient	High	Low	Medium
Stuck in local minimum	Easy	Difficult	Difficult
Easy to converge	Yes	No	No

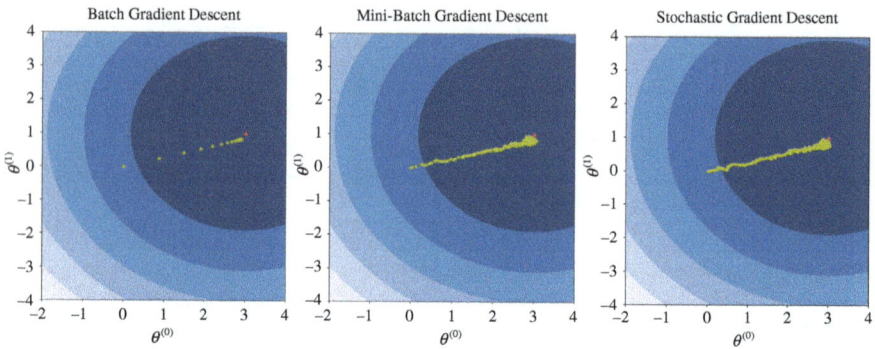

Figure 2.4. Convergence of parameters $(\theta_n)_n$ for BGD, mini-batch GD and SGD.

2.1.4 Prediction and validation

There are various ways of judging goodness of fit, which can be mainly divided into two types:

- statistics-based approach;
- machine learning-based approach.

2.1.4.1 *Statistics-based approach*

For statistics based validation, we usually need to make extra probabilistic assumptions of the residuals. Hypothesis testing is a hypothesis that is testable on the basis of observing a process that is modeled via a set of random variables—e.g., p-value, R^2, R^2_{adj}.

- **p-value:** Under the null hypothesis, the probability that the statistical summary is equal to or more extreme than the observed one. A smaller p-value indicates rejecting the null hypothesis. However, it does not measure the probability of making mistakes by rejecting a true null hypothesis (a Type I error).

- R^2: The proportion of the variance in the output variable that is predictable from input variables, also called the coefficient of determination.

$$R^2 = 1 - \frac{\sum_{i=1}^{N}(y_i - x_i^T \hat{\beta})^2}{\sum_{i=1}^{N}(y_i - \bar{y})^2}, \tag{2.14}$$

where $\hat{\beta}$ is the optimal parameter in the linear model.

- **Adjusted R^2**: A similar concept to R^2 that takes the numbers of model parameters (input dimension) into account. It is defined in the following form, which penalizes larger input dimensions:

$$R_{adj}^2 = 1 - (1 - R^2)\frac{N-1}{N-d-1}, \tag{2.15}$$

where $\hat{\beta}$ is the optimal parameter in the linear model and d is the input dimension.

2.1.4.2 *Machine learning-based approach*

Machine learning-based validation focuses mainly on predictive power on an unseen new dataset, which is the generalization ability of the fitting model. To achieve this, one usually divides the dataset into a training dataset and a testing dataset. The model is calibrated using the training dataset, and the goodness of fit is computed for both the training set and testing set.

Perfect fitting on the training set is usually not a good thing because typically, the training set contains some random noise; this noise should be filtered out to give good generalization. For example, if you choose an over-complicated model which includes too many parameters, the model might have a perfect performance on the training set, as it mistakenly regards the noise as part of the signal of the model. This model may then not perform well on the testing set as there will be new, different noise, leading to little predictive power. This kind of problem is called the *overfitting* issue as the model overfits the training data. This is a very common but important problem in the training process.

On the other hand, if you choose a simple model, it may be not rich enough to describe the complex relationship between inputs and outputs. This is called the *underfitting* issue. An illustration of fitting issues in the training process is shown in Figure 2.5.

Commonly used indicators for goodness of fit include the mean squared error (MSE), R^2 and R_{adj}^2.

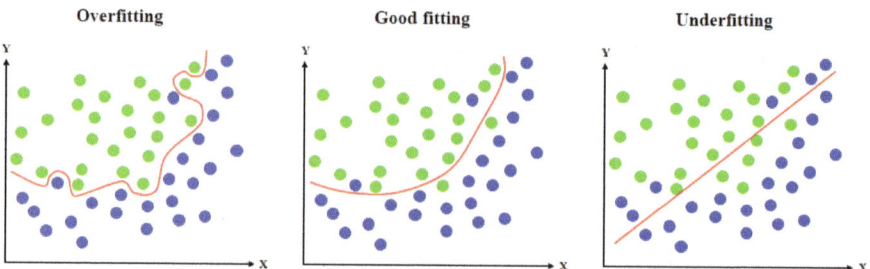

Figure 2.5. Potential fitting issues in the training process.

2.1.4.3   *Cross-validation and parameter tuning*

As we discussed earlier, the ultimate goal of supervised learning is to train a model using labeled data, which can be generalized to an unseen dataset. Thus the evaluation of the predictive power of a model is crucial. This is typically assessed by *cross-validation* in practice.

The main idea of model assessment is to further split the training data into two parts, i.e., a subset of training data for training the model and a validation set to assess the predictive power of the trained model (without touching the test data). We choose the model that achieves the best performance measure in the validation set as the final model and use this model to predict in the testing set. However, this may drastically reduce the size of the training set. A solution to this problem, called k-fold cross-validation (Figure 2.6), is to split the training set into k subsets ("folds") and conduct the following procedures:

(1) Train a model using $(k-1)$ folds of the training data.
(2) Use the remaining fold for the validation to compute the performance measurement (e.g., MSE) of this model.

For one model with a given set of hyperparameters, we conduct k-fold cross-validation, and the average performance measure of the k folds can be used as a scalar score to measure its performance. This can be combined with *grid search* to select optimal hyperparameters. As its name suggests, grid search is an exhaustive search method of choosing the best hyperparameters. It requires pre-specifying possible values of a hyperparameter (grid) and choosing the best one based on the corresponding cross-validation scores. Lastly, once the optimal hyperparameters are chosen, we refine

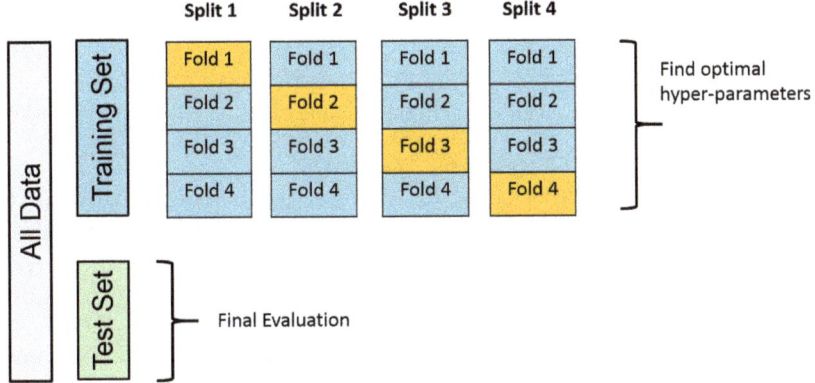

Figure 2.6. 4-fold cross-validation and parameter tuning.

the model using the whole training set and make the prediction in the testing set.[5]

Note that cross-validation and grid search are standard methods for the performance measurement and parameter tuning of both regressors and classifiers. Thus in the next section on validation of classification, we skip the discussion on cross-validation and grid search.

2.2   From Regression to Classification

2.2.1   Categorical output

The setup for classification is very similar to that of regression problems. Given a set of input–output pairs $\mathcal{D} = \{(x_i, y_i)\}_{i=1}^N$, we aim to infer the functional relationship between an input x and an output y. But classification differs from regression mainly because the output variable of the classification is categorical. In other words, there are only finite many possible values of y_i, denoted by $\mathcal{Y}$. W.l.o.g., $\mathcal{Y} = \{1, \cdots, n_o\}$, where n_o denotes the number of possible categories. According to the different numbers of possible categories, classification can be divided into binary classification ($n_o = 2$) and multi-class classification ($n_o > 2$).

Categorical variables represent a qualitative method of scoring data (i.e., they represent categories or group membership). For example, the

---

[5]Interested readers may refer to `https://scikit-learn.org/stable/modules/cross_validation.html` for more details of cross-validation and its implementation in the Scikit-Learn package.

Table 2.10. Different encoding methods for the blood type example.

Blood type	A	B	AB	O
Integer encoding	1	2	3	4
One-hot vector encoding	0001	0010	0100	1000

blood type of a person may be A, B, AB or O, which is a categorical variable. There are several ways to represent categorical variables numerically, including

- integer encoding (the ith class is represented using an integer i);
- one-hot vector encoding (the ith class is represented using a binary vector of length n_o, which has the unique non-zero element at the ith position).

Let us revisit the example of blood type. The numerical representation of blood types using two above encoding methods are given in Table 2.10.

2.2.2   Model

The objective of classification is to predict the corresponding output for any given new input x_*, just like the regression problem. However, due to the categorical nature of the output, this question has a slightly different mathematical formulation from that of regression problems. In the classification problem, instead of predicting the output y directly, we aim to estimate the probability of the output being y conditional on an input x, which is described by a model $f_\theta : E \to \mathbb{R}^{n_o}$. Intuitively, we have that

$$\langle f_\theta(x), \bar{y} \rangle \approx \mathbb{P}[y|x],$$

where $\bar{y}$ is one-hot encoding of the class y, and $\langle .,. \rangle$ is the inner product of two vectors of length n_o.

Let us first understand why we do not aim to predict the conditional expectation of the output as we do in regression. This is because the conditional expectation of a categorical output does not make sense. For example, if the conditional distribution of the output label is known as a discrete random variable with probability $(0.4, 0.2, 0.4)$, the conditional mean of the output depends on the numeric representation of the output category. More importantly, in classification, one can't infer the best estimator for the output category given the input based on only this conditional mean of output. Therefore we usually estimate the conditional probability of each class label.

2.2.3 Loss function and optimization

In contrast to the quadratic loss function in regression, the cross entropy loss function is commonly used in classification as it provides a way to quantify the difference between the empirical conditional distribution of output y given the input x and the model estimated conditional distribution $f_\theta(x)$. For discrete probability distributions p and q with the same support $\mathcal{Y}$, the cross entropy is defined to be

$$H(p,q) := -\sum_{j\in\mathcal{Y}} p(j)\log(q(j)).$$

For a given distribution p, H is a function of q, and it attains its smallest value when $q = p$. Intuitively, smaller cross entropy $H(p,q)$ means that two distributions are similar. In other words, when minimizing the cross entropy H, the optimal distribution of q is the same as that of p.

Definition 2.3 (Cross Entropy Loss Function). The cross entropy loss function $Q_\theta : E \times \mathcal{Y} \to \mathbb{R}$ is defined to be

$$Q_\theta(x,y) = -\langle y, \log f_\theta(x)\rangle,$$

where $x \in E$, y is one-hot encoding in $\mathcal{Y}$, θ are model parameters of f_θ, and $\langle .,.\rangle$ is the inner product.

The corresponding empirical cross entropy loss function is given as the average of the above cross entropy loss function evaluated at all samples:

$$L(\theta|\mathcal{D}) = -\frac{1}{N}\sum_{i=1}^{N}\langle y_i, \log f_\theta(x_i)\rangle.$$

Another way to interpret the cross entropy is through maximum likelihood estimation (MLE). The cross entropy loss function can be regarded as the negative log-likelihood function of θ, given the observation of the input–output pairs. Assuming all the samples are mutually independent, the likelihood function is given as the product of the conditional probability of the output, i.e.,

$$\prod_{i=1}^{N}\langle f_\theta(x_i), \bar{y}_i\rangle \to \max,$$

which is equivalent to minimizing the negative log-likelihood ratio, i.e.,

$$-\sum_{i=1}^{N}(\langle\log f_\theta(x_i), \bar{y}_i\rangle) \to \min.$$

That is exactly the cross entropy loss function, denoted by $L(\theta|\mathcal{D})$ (see Definition 2.3). Thus it is noted that the optimal parameters from minimizing the cross entropy are the same as those obtained by maximizing the likelihood. The cross entropy empirical loss has an additive form, which allows parallel computation benefit for each sample.

In the next stage of the optimization to find the optimal parameter θ^* to minimize $L(\theta|\mathcal{D})$, we usually make the further assumption that f_θ is differentiable w.r.t. θ. The numerical optimization methods we have discussed in Section 2.1.3 can be exploited here as well.

2.2.4   Prediction and validation

Once we obtain the optimal parameters θ^*, the prediction is straightforward. For any new input data x_*, use the output label with the highest estimated conditional probability as the estimator for the output,

$$\hat{y}_* = \arg\max_{i \in \mathcal{Y}} f_{\theta^*}^{(i)}(x_*),$$

where $f_{\theta^*}^{(i)}(x_*)$ is the ith coordinate of $f_{\theta^*}(x_*)$.

At the final stage, we need to specify the metric of the goodness of fit. There are various performance measures, e.g., the accuracy, the confusion matrix, etc.

2.2.4.1   *Accuracy*

In classification, the dimension of the model output $f_{\theta^*}(x)$ is n_o, which represents the estimated conditional probability of each output. Let $\hat{Y}_{\text{prob}}$ denote the matrix of size (N, n_o),

$$\hat{Y}_{\text{prob}} = (f_{\theta^*}(x_i))_{i \in \{1,2,...,N\}}. \tag{2.16}$$

For multi-class classification, the accuracy is one of the most popular measures and is defined as follows:

$$\sum_{i=1}^{N} \frac{\mathbf{1}(\hat{y}_i = y_i)}{N},$$

where $i \in \{1, 2, \cdots, N\}$, and y_i and $\hat{y}_i$ denote the actual output and the estimated output of the ith sample, respectively.

2.2.4.2 *Confusion matrix*

Another way to measure the performance of a classifier is the confusion matrix. The column represents the estimated label for the classification problem and the row represents the true label. Let $M := (M_{i,j})_{i,j \in \mathcal{Y}}$ denote the confusion matrix, where $M_{i,j}$ denotes the number of samples with true label i and estimated label j. The better the prediction, the more diagonally dominant the confusion matrix M is.

The normalized confusion matrix is defined from the confusion matrix and denoted by $\hat{M} = (\hat{M}_{i,j})_{i,j \in \mathcal{Y}}$, where $\hat{M}_{i,j}$ is defined as follows:

$$\hat{M}_{i,j} = \frac{M_{i,j}}{\sum_{j \in \mathcal{Y}} M_{i,j}}.$$

$\hat{M}_{i,j}$ represents the empirical conditional probability of the sample being identified as j when it in fact belongs to class i. The better the prediction, the closer $\hat{M}$ is to the identity matrix.

2.2.4.3 *Other metrics for binary classification*

You may wonder why we need other metrics than accuracy to assess classification performance. When the data are extremely imbalanced, the trivial classifier (estimating all samples as the majority class) gives very high accuracy, which implies that accuracy is not an informative performance measure in this case. Next, we introduce some other commonly used metrics for the binary classification case: precision, recall, PR curve and ROC curve.

As shown in Figure 2.7, the confusion matrix of a binary classifier $M = (M_{j_1,j_2})_{j_1,j_2 \in \{1,2\}}$ is a 2×2 matrix, where

- True Positive (TP, $M_{2,2}$): the number of samples that have actual label class 2 and predicted label class 2.
- False Positive (FP, $M_{1,2}$): the number of samples that have actual label class 1 and predicted label class 2.

Predicted Label / Actual Label	Class 1 (Negative)	Class 2 (Positive)
Class 1(Negative)	$M_{1,1}$ (TN)	$M_{1,2}$(FP)
Class 2(Positive)	$M_{2,1}$ (FN)	$M_{2,2}$(TP)

Figure 2.7. Confusion matrix of a binary classifier.

- True Negative (TN, $M_{1,1}$): the number of samples that have actual label class 1 and predicted label class 1.
- False Negative (FN, $M_{2,1}$): the number of samples that have actual label class 2 and predicted label class 1.

The *precision* of a binary classifier is defined as the percentage of true positive samples among all the samples with a predicted label of "positive":

$$\text{precision} = \frac{\text{TP}}{\text{TP} + \text{FP}}.$$

The *recall*, also called the *sensitivity* or *true positive ratio* (TPR), is defined as the percentage of true positive samples among all the samples with an actual label of "positive":

$$\text{recall} = \frac{\text{TP}}{\text{TP} + \text{FN}}.$$

From the above definition, we can see that one trivial way to get a high recall is to predict all the samples being "positive," which gives perfect recall of 100%. Thus the recall is typically used accompanied by the precision. A higher recall suggests a larger TP, which indicates a better performance of the classifier. But increasing the recall reduces the precision and vice versa. This is called the precision and recall trade-off.

Next, let us introduce the *precision–recall curve* (PR curve) based on the above concepts of precision and recall. For a classifier, f_{θ^*} gives an estimated probability (also called a score) to each possible output class. Instead of choosing the class label that gives the maximum score, alternatively for each given threshold value t, we assign the estimator for the output using the following equation:

$$\hat{y} = \begin{cases} 1, & \text{if } f_{\theta^*}(x) > t; \\ 2, & \text{if } f_{\theta^*}(x) \leq t. \end{cases} \tag{2.17}$$

Varying the threshold t, the corresponding precision and recall can be computed, and thus the PR curve is obtained.

The *receiver operating characteristic curve* (ROC curve) is another important metric of a binary classifier. Similar to the PR curve, varying the threshold t, the ROC curve is the curve of TP against FP. AUC stands for "area under the ROC curve." It measures the entire two-dimensional area enclosed by the ROC curve, a line from $(0,0)$ to $(1,0)$ and a line from $(1,0)$ to $(1,1)$.

2.2.4.4 *Numerical example*

In the following, we use the binary classification of identifying whether a digit image is a number 8 as a concrete example to show how to compute all the metrics we have discussed and implemented it using Scikit-Learn.

We use the MNIST dataset composed of digit images of the numbers 0–9.[6] The input data is a gray-valued image, and the output is the digit in the input image. Now we want to identify whether an input image is a digit 8 and construct a binary classifier where class 1 represents "non-8 digit" while class 2 represents "8 digit." The training dataset contains 60,000 handwritten digit images, including 54,149 non-8 digit samples and 5851 8 digit samples. It is easy to see that there are many more negative class cases than those of the positive class. Thus this is a class imbalance problem. Figure 2.8(left) shows the estimated output (score) of the first 15 samples, and each row has the sum 1. The second column represents the estimated probability of the class label being 2. The estimated class label is the label with the maximum probability, and Figure 2.8(right) provides the estimated output label.

conditional probability estimator			estimated output	
	class 1 (non 8)	class 2 (digit 8)		class 1 (non 8)
0	0.999929	0.000071	0	False
1	0.999096	0.000904	1	False
2	0.999970	0.000030	2	False
3	0.877981	0.122019	3	False
4	0.999048	0.000952	4	False
5	0.982514	0.017486	5	False
6	0.993879	0.006121	6	False
7	0.918145	0.081855	7	False
8	0.962651	0.037349	8	False
9	0.993038	0.006962	9	False
10	0.995717	0.004283	10	False
11	0.345118	0.654882	11	True
12	0.999995	0.000005	12	False
13	0.999986	0.000014	13	False
14	0.931200	0.068800	14	False

Figure 2.8. The estimated output of the first 15 samples.

---

[6]http://yann.lecun.com/exdb/mnist/.

Code:

```
1   from sklearn.metrics import confusion_matrix
2   # Y_test is a binary vector of the actual class label with dim (N, 1) where
    ↪   N is the number of samples;
3   # y_test_est is a binary vector of the estimated class label with dim (N,
    ↪   1).
4
5   cm = confusion_matrix(Y_test, y_test_est)
6   print('confusion matrix is {}'.format(cm))
```

Screen Output:

confusion matrix is

$$\begin{pmatrix} 8810, \ 216 \\ 319, \ 655 \end{pmatrix}.$$

Figure 2.9. Python code for computing the confusion matrix and the corresponding result.

We first compute the confusion matrix on the test set using confusion_matrix() in the Scikit-Learn package as shown in Figure 2.9.

Based on the confusion matrix, we can compute the corresponding accuracy via

$$\text{accuracy} = \frac{\text{TP} + \text{TN}}{\text{TP} + \text{FN} + \text{TN} + \text{FP}} = \frac{8810 + 655}{10000} = 0.9465.$$

You may also use accuracy_score() in sklearn.metrics to compute the accuracy of $\hat{Y}$:

```
1   from sklearn.metrics import accuracy_score
2   acc = accuracy_score(Y_test, y_test_est)
```

The accuracy is about 94.65%, which seems very good. But if one computes the precision and recall, the prediction is not that great.

$$\text{precision} = \frac{\text{TP}}{\text{TP} + \text{FP}} = \frac{655}{216 + 655} = 0.7520$$

$$\text{recall} = \frac{\text{TP}}{\text{TP} + \text{FN}} = \frac{655}{319 + 655} = 0.6725.$$

Similar to the accuracy case, you may use the following Python function to compute the precision and recall:

Code:

```
1  from sklearn.metrics import precision_recall_curve
2  precisions, recalls, thresholds = precision_recall_curve(Y_test,
   ↪  y_test_prob_est[:,1])
3  plt.plot(precisions, recalls, 'b')
4  plt.xlabel('precision', fontsize=14)
5  plt.ylabel('recall', fontsize=14)
6  plt.title('PR Curve')
7  plt.axis([0, 1, 0, 1])
```

Screen Output:

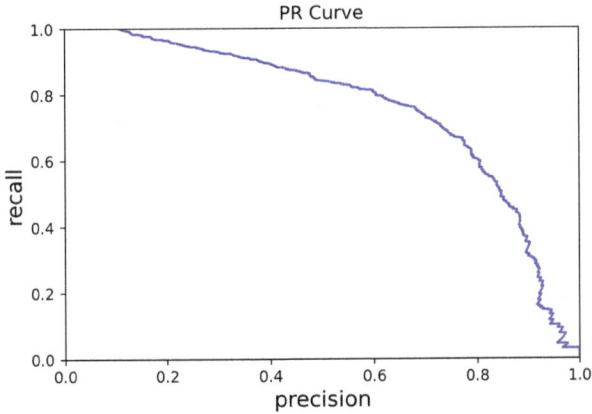

Figure 2.10. The code for the PR curve plot and the screen output.

```
1  from sklearn.metrics import precision_score, recall_score
2  precision = precision_score(Y_test, Y_test_est)
3  recall = recall_score(Y_test, y_test_est)
```

Figures 2.10 and 2.11 provide the code for computing the PR curve and ROC curve obtained by a binary classification task using the Scikit-Learn Python package. In this example, the AUC score is 0.9423, which is the area of the blue shaded region enclosed under the ROC curve.

Remark 2.1. When the positive class has much fewer samples than the negative class and the false positives are more important, one should choose the PR curve. For example, looking at the previous ROC curve and the AUC score, you may think that the classifier is really good. But this is mostly because there are few positives compared to the negatives.

Code:

```
1   from sklearn.metrics import roc_curve, roc_auc_score
2   fps, tps, thresholds = roc_curve(Y_test, y_test_prob_est[:,1])
3   roc_auc_score_train = roc_auc_score(Y_test, y_test_prob_est[:,1])
4   import matplotlib.pyplot as plt
5   plt.plot(fps, tps, 'b')
6   plt.xlabel('false positive rate', fontsize=14)
7   plt.ylabel('true positive rate', fontsize=14)
8   plt.title('ROC Curve')
9   plt.axis([0, 1, 0, 1])
10  plt.fill_between(fps, 0, tps, facecolor='lightblue', alpha=0.5)
11  plt.text( 0.5, 0.8, 'roc auc score = '+str(round(roc_auc_score_train, 4)),
    ↪   fontsize=14)
12  plt.annotate("",
13              xy=(0.3, 0.7), xycoords='data',
14              xytext=(0.5, 0.8), textcoords='data',
15              arrowprops=dict(arrowstyle="->",
16                              connectionstyle="arc3"),)
```

Screen Output:

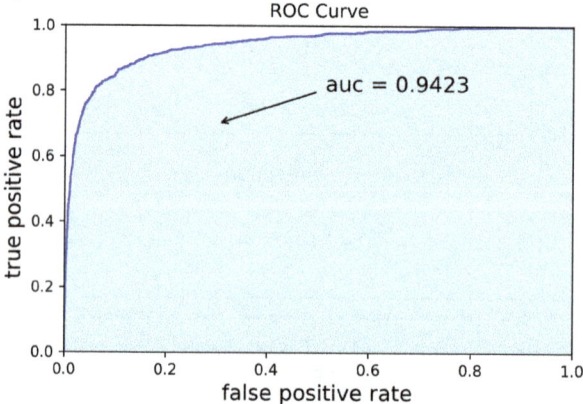

Figure 2.11. The ROC curve.

In conclusion, Table 2.11 provides a summary of the general framework of classification.

2.3   Model Ensemble

As the saying goes, two heads are better than one. There exists a similar principle in machine learning. One may wonder whether aggregating

Table 2.11. The framework of classification.

Dataset:	$\mathcal{D} = \{(x_i, y_i)\}_{i=1}^N$.
Model:	$f_\theta(x, y) \approx P(y\|x)$, $\forall x \in \mathbb{R}^d$, $y \in \mathcal{Y}$.
Empirical Loss:	$L(\theta\|\mathcal{D}) = -\frac{1}{N} \sum_{i=1}^N \log(f_\theta(x_i, y_i)) \to \min$.
Optimization:	$\theta^* = \arg\min_\theta(L(\theta\|\mathcal{D}))$.
Prediction:	$\hat{y}_* = \arg\max_{y \in \mathcal{Y}} f_{\theta^*}(x_*, y)$.
Validation:	Accuracy, confusion matrix, etc.

different predictors can have better prediction performance than could be obtained from any of the constituent learning algorithms alone. The answer is yes for most cases. Ensemble learning is devoted to addressing this question.

2.3.1 Intuition of ensemble

An ensemble is nothing other than a collection of predictors that are combined together (e.g., the majority of all predictions) to give a final prediction. The reason that we use the ensemble method is that one can incorporate many predictors of the same output variable to improve the prediction performance over that of any single predictor.

We use the following simple numerical example to illustrate the idea behind ensemble methods. Assume that there is a binary classification problem with all sample labels being 2. Now say we only have a classifier with 55% accuracy; this means that it predicts correct class labels with probability 0.55. We simulate this classifier in Listing 2.1. Obviously, it is a weak learner as it only performs a little bit better than random guessing. So what should we do to improve the performance without the help of new learners? The ensemble method can help us out. We can simply combine multiple (e.g., 1000) identical weak learners and use majority voting to decide the estimated class for the 1000 learners. If most learners return 2, then the estimated class of the ensemble model is correct. We simulate 10,000 samples, and the accuracy of one weak learner is 55%, which is close to the setting of the problem. We assume that the weak learners are mutually independent. You will find that an ensemble model with 1000 weak learners should achieve nearly 100% accuracy, which is an amazing result. The accuracy of ensemble models with different numbers of learners is depicted in Figure 2.12. The accuracy gradually increases with an increasing number of learners.

```
1    # importing mean()
2    from statistics import mean
3    def weak_learner():
4        n = np.random.randint(0, 100)
5        return 1 if n >= 45 else 0
6
7    # Majority voting method
8    def majority_voting(results:list):
9        return 1 if results.count(1) > results.count(0) else 0
10
11   # Define ensemble model with 1000 weak learners
12   def ensemble_model(learner, num_learners = 1000):
13       all_results = [learner() for i in range(num_learners)]
14       return majority_voting(all_results)
15
16   # Simulate 10,000 samples to approximate the accuracy
17   num_samples = 10000
18   all_weak_learner_results = []
19   all_ensemble_model_results = []
20   for i in range(num_samples):
21       weak_learner_result = weak_learner()
22       ensemble_model_result = ensemble_model(weak_learner)
23       all_weak_learner_results.append(weak_learner_result)
24       all_ensemble_model_results.append(ensemble_model_result)
25
26   print('The weak learner only achieves accuracy
      ↪    of',mean(all_weak_learner_results))
27   print('The ensemble model achieves accuracy as high as',
      ↪    mean(all_ensemble_model_results))
```

Listing 2.1. Python code for a numerical example of ensemble methods.

As we can see from the above example, the ensemble method turns the weak learner into a strong learner, and the accuracy improves sharply. In practice, we have to face more complicated problems, e.g., multi-classification and regression problems. Though the ensemble method may not perform as amazingly as in the above example, it is the most popular method for model selection.

2.3.2 Homogeneous weak learners ensemble
In the following, we divide model ensemble methods into two types based on types of weak learners:

- Homogeneous weak learners (the base models have the same type):
  - Bagging/Pasting
  - Boosting

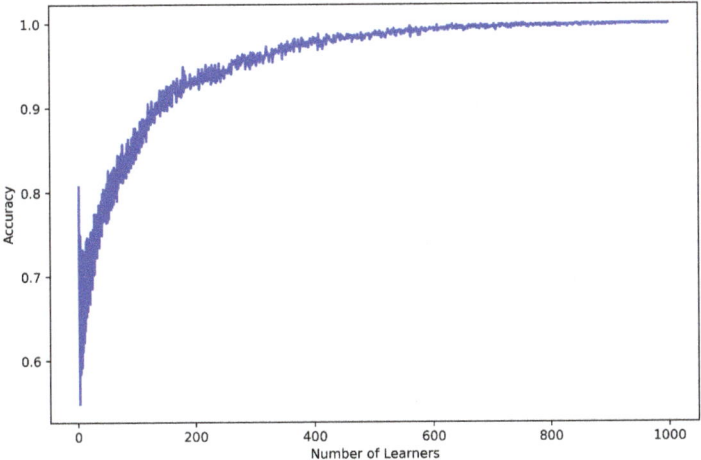

Figure 2.12. The accuracy of ensemble models with different numbers of learners.

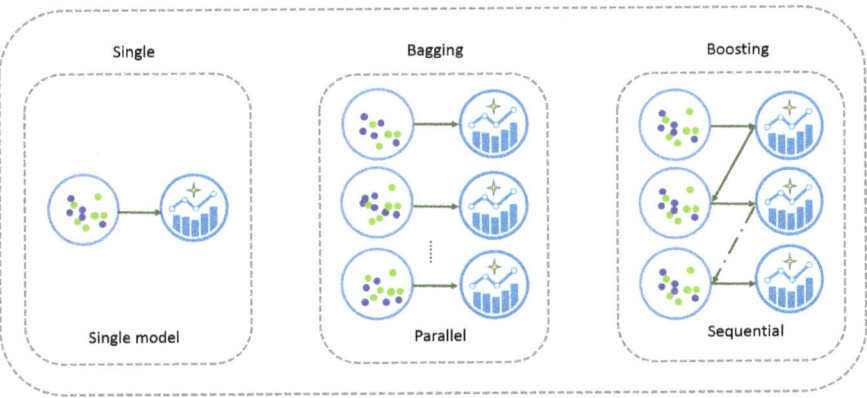

Figure 2.13. Homogeneous weak learners ensemble methods.

- Heterogeneous weak learners (the base models may have different types):
  - Stacking

As we can see in Figure 2.13, ensemble techniques of homogeneous weak learners are further classified into the following main types:

(1) To use the same training algorithm for predictors, but each time a subset of samples are randomly selected for training. In this case, we typically combine predictors using some model averaging techniques, e.g., weighted average, majority vote or normal average.

(2) To combine the predictors into the final predictor in a sequential manner, which is called boosting.

The first type of ensemble method usually involves aggregating many uncorrelated learners, which reduces error by reducing variance. Under this category, Bagging (short for bootstrap aggregating) [Breiman (1996)] and Pasting [Breiman (1999)] are the two major sub-classes. For Bagging, each observation is chosen with replacement to be used as input for each of the model. In contrast, for Pasting, each time a subset of data is randomly selected without replacement.

In the following, we focus on out-of-folds (OOF), which is another model ensemble method that falls into this category. OOF refers to a step in the learning process when using k-fold cross-validation in which the predictions from each set of folds are grouped into predictions of the training set. These predictions are now "out-of-folds," and thus the error can be calculated on these to get a good measure of how good your model is. As shown in Figure 2.14, the procedure is composed of the following steps, and the algorithm is outlined in Algorithm 1.

(1) Split the dataset into the training and testing set.
(2) Use stratified k-fold cross-validation in the training set and thus obtain k estimated models.
(3) Evaluate each estimated model on the testing set and therefore have k estimators of the testing data.

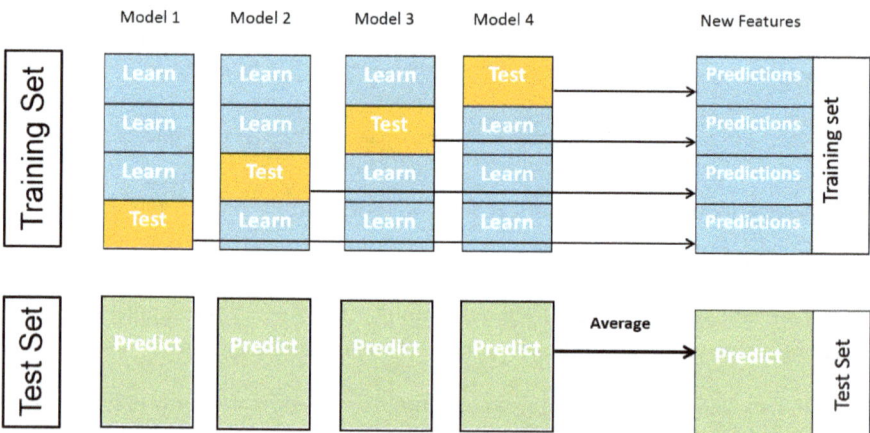

Figure 2.14. Illustration of OOF prediction procedure.

Algorithm 1: OOF Prediction Algorithm

1: **Input:** $\mathcal{D} = (X_{train}, Y_{train})$, X_{test}, K

2: Split the training set into K folds, denoted as $\mathcal{D}_1, \mathcal{D}, \ldots, \mathcal{D}_K$;

3: Set $\hat{Y}_{test} = 0$;

4: **for** $i = 1 : K$ **do**

5:     Train the model in $\mathcal{D}/\mathcal{D}_i$ and obtain a model T_i;

6:     Calculate the predictor of the testing data using model T_i, denoted by $\hat{Y}^{(i)}$;

7:

$$\hat{Y}_{test} = \hat{Y}_{test} + \hat{Y}^{(i)}.$$

8: **end for**

9: The final estimator of $\hat{Y}_{test}$ is given as follows:

$$\hat{Y}_{test} = \frac{\hat{Y}_{test}}{K}.$$

10: **Output:** $\hat{Y}_{test}$.

---

(4) Average the k estimators of the testing data to get the final estimator of the testing data.

For the second type of ensemble method, the core idea of boosting is to update subsequent predictors based on the error of the previous predictors. Because new predictors are updated from learning mistakes by previous predictors, it takes fewer iterations to get close to ground-truth predictions. But the stopping criterion is essential in this case. If it is not appropriately chosen, it could easily lead to overfitting on training data.

Gradient boosting is an example of a boosting algorithm.[7] We devote the rest of this subsection to explain the gradient boosting method. It is a variant of the gradient descent algorithm that provides a way to combine weaker learners to get better estimation for the gradients and construct a final learner in order to provide better prediction.

Recall that the objective is to minimize the loss function $L(\theta|(X,Y))$. In gradient boosting, the weak learner (h_m) is used as a base model, and then a sequence of predictors $(f_m)_{m=1}^{M}$ is constructed, where the updating

---

[7]https://medium.com/mlreview/gradient-boosting-from-scratch-1e317ae4587d.

rule of f_m is given as follows:

$$f_m = f_{m-1} + \gamma_m h_m, \tag{2.18}$$

where $h_m \in \mathcal{H}$, which is the set of base models, and $\gamma_m \in \mathbb{R}$ is a constant. In this case, f_m is an additive model: when m is increased by 1, the parameter of the model h_m is added to the parameters of f_m.

Compare the weight update rule of the gradient descent algorithm and Equation (2.18): h_m should serves as the gradient term $\nabla L(\theta|(X,Y)) := \nabla L(Y, f_M(X))$, where θ is the set of all parameters of f_M. However, at the mth iteration, we cannot evaluate $\nabla L(\theta|(X,Y))$ as $(h_j)_{j=m}^{M}$ are unknown. It is natural to use $\nabla L(Y, f_{m-1}(X))$ to approximate the actual gradient. But $\nabla L(Y, f_{m-1}(X))$ may be noisy and it may not belong to any base model. Therefore this suggests using the base model h_m to fit the derivative terms $\nabla L(Y, f_{m-1}(X))$. Thus the update rule of the gradient boosting algorithm is proposed: at each mth iteration we update f_m using

$$f_m(x) = f_{m-1}(x) - \gamma_m \nabla_{f_{m-1}} L(y, f_{m-1}(x)). \tag{2.19}$$

γ_m can be chosen by solving the following one dimensional optimization problem:

$$\gamma_m = \arg\min_{\gamma} L(y - f_{m-1}(x) - \gamma \nabla_{f_{m-1}} L(y, f_{m-1}(x)).$$

Then the gradient boosting algorithm is given in Algorithm 2.

Let us consider the regression problem, which aims to minimize the quadratic loss function $L(\theta|(X,Y))$. Then the derivative term can be simplified to residuals as follows:

$$\nabla_f L(y, f(x)) = 2(y - f(x)).$$

In this case, h_m can be viewed as correcting the error terms by learning the residuals of the previous estimator f_{m-1}.

2.3.3   Heterogeneous weak learners ensemble

Stacking is a heterogeneous ensemble method to build a meta-model using predictors from various models. The main idea is to use the predictors of each model as new inputs and learn the relationship between the model predictors and the output.

As shown in Figure 2.15, the procedure of this type of model stack is outlined as follows:

(1) Split the dataset into the training and testing set.

Algorithm 2: Gradient Boosting Algorithm

1: **Input:** $(x_i, y_i)_{i=1}^N$.

2: Initialize f_0 by a constant γ_0 via the following equation:

$$\gamma_0 = \arg\min_{\gamma} L(y, \gamma);$$

3: **for** $m = 1 : M$ **do**

4:     **for** $i = 1 : N$ **do**

5:         Compute the residuals

$$r_{im} = \left[\frac{\partial L(y_i, f(x_i))}{\partial f(x_i)} \right]_{f=f_{m-1}}.$$

6:     **end for**

7:     Fit a base model learner h_m to the target r_{im}, using the data $(x_i, r_{im})_{i=1}^n$.

8:     Solve the one dimensional optimization problem

$$\gamma_m = \arg\min_{\gamma} \sum_{i=1}^n L(y_i, f_{m-1}(x_i) + \gamma h_m(x_i)).$$

9:     Update f_m using the following formula:

$$f_m(x) = f_{m-1}(x) + r_m h_m(x).$$

10: **end for**

11: **Output:** f_M.

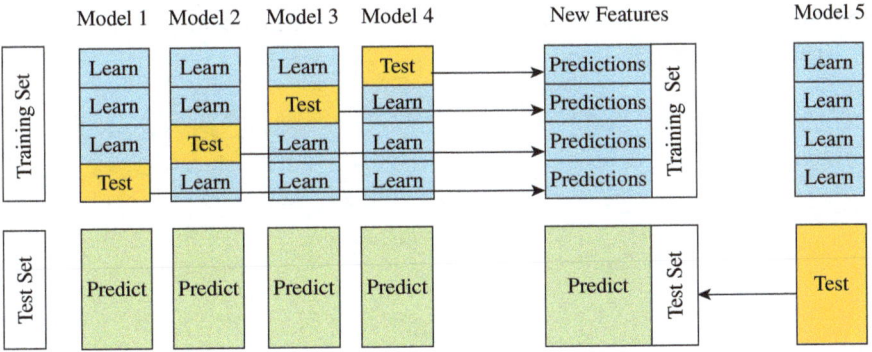

Figure 2.15. Illustration of stacking.

Table 2.12. Summary of three model ensemble methods.

- *Bagging*, often built on top of homogeneous weak learners by randomly selecting the subset for the training set with replacement and combining them by a deterministic averaging/voting process.

- *Boosting*, often built on top of homogeneous weak learners, which is constructed in a sequential and adaptive way (a base model depends on the previous ones) and combining them by a deterministic averaging/voting process.

- *Stacking*, often built on top of heterogeneous weak learners, which is constructed in parallel and combines them by training a meta-model to output a prediction based on the different weak learners.

(2) Use stratified k-fold cross-validation on the training set. Each sample in the training set appears only once in the validation set of the k-fold cross-validation. We add the predicted output as a new feature.
(3) Repeat above step for n models.
(4) Learn a new meta-model using the new features (and optionally the original input) as the input and output on the training set.
(5) Make a prediction using the meta-model on the testing data.

In conclusion, Table 2.12 gives a summary of the three main types of model ensemble method we have discussed.

2.4   Exercises

(1) What is the supervised learning problem?
(2) Is the forecasting of the future price of some stock a regression problem?
(3) What is the commonly used loss function in the regression problem?
(4) What is the cross entropy?
(5) What is a categorical variable?
(6) What is the difference between the regression problem and the classification problem?

Chapter 3

Linear Regression and Regularization

In this chapter, we focus on linear regression and explain how it fits in the general framework of supervised learning. We start with the simplest linear model—Ordinary Least Squares (OLS)—and then discuss the natural extension of OLS, i.e., the linear model with regularization.

3.1 Ordinary Least Squares Method

3.1.1 Derivation

Ordinary Least Squares (OLS) is the simplest linear model in regression, and it has been widely used in many diverse fields, e.g., economics, political science, and engineering. We follow the general setup of the regression problem in Section 2.1. OLS assumes that the model between the input $x \in \mathbb{R}^d$ and the output $y \in \mathbb{R}$ is linear, and as a formula,

$$y = x\theta + \varepsilon, \tag{3.1}$$

where x is a d-dimensional row vector, θ is a d-dimensional column vector, and ε is a scalar noise term with conditional mean on the input x being zero. θ represents the fixed but unknown linear coefficients of the OLS model. As the name OLS suggests, the loss function of OLS is the sum of the squared residuals

$$L(\theta|X, Y) = \sum_{i=1}^{N} (y_i - x_i\theta)^2 = (Y - X\theta)^T (Y - X\theta),$$

where $(x_i, y_i)_{i=1}^{N}$ is a collection of the input–output pairs and (X, Y) is defined as in Equation (2.2).

Remark 3.1. For the case where the linear model includes a non-zero intercept, i.e.,

$$y = \theta_0 + x\theta + \varepsilon, \tag{3.2}$$

we reformulate it as the input $\tilde{x} = (1, x)$ by adding constant 1 to an extra coordinate of input variable x, and then Equation (3.2) can be rewritten as

$$y = \tilde{x}\tilde{\theta} + \varepsilon, \tag{3.3}$$

where $\tilde{\theta} = \begin{pmatrix} \theta_0 \\ \theta \end{pmatrix}$.

The loss function can be connected with the maximum log-likelihood estimator of the linear model by the assumption of the normal distribution of the residuals. Note that if one assumes identical and independent distributions of the sample noise residuals $(\varepsilon_i)_{i=1}^N$ with zero conditional mean and constant conditional variance, minimizing the loss function of OLS is equivalent to maximizing the log-likelihood ratio of the observation of the output $(y_i)_{i=1}^N$ conditional on $(x_i)_{i=1}^N$.

One great advantage of OLS is that it yields an analytic formula for the optimal parameter estimator for θ, denoted by $\hat{\theta}$. By matrix computation, one may easily obtain the following formulae for $\hat{\theta}$ (Equation (3.4) in Lemma 3.1).

Lemma 3.1 (OLS Estimator). *Given the input–output data set $\mathcal{D} = (X, Y)$, and assuming the standard setting of the OLS model holds and the existence of the inverse of $X^T X$, the estimator of the optimal parameter $\hat{\theta}$ is given as follows:*

$$\hat{\theta} = (X^T X)^{-1} X^T Y. \tag{3.4}$$

Proof. Recall that the loss function of OLS yields that

$$\begin{aligned} L(\theta|X, Y) &= (Y - X\theta)^T (Y - X\theta) \\ &= Y^T Y - 2\theta^T X^T Y + \theta^T X^T X\theta. \end{aligned} \tag{3.5}$$

We use the fact that $Y^T X\theta = (X\theta)^T Y$ as both sides are scalars and the transpose of a scalar remains unchanged. It is noted that the loss function of OLS is a quadratic function with respect to the parameter θ, and it is thus a convex function, which ensures the uniqueness and existence of the global minimum of the optimal parameters. Besides, $L(\theta|X, Y)$ is differentiable with respect to θ, and thus the optimal parameter $\hat{\theta}$ should satisfy that the derivative of $L(\theta|X, Y)$ evaluated at $\theta = \hat{\theta}$ is equal to zero, i.e.,

$$\frac{\partial L(\theta|X, Y)}{\partial \theta}\Big|_{\theta=\hat{\theta}} = 0.$$

By Equation (3.4), it follows that

$$\frac{\partial L(\theta|X, Y)}{\partial \theta} = -2X^T Y + 2X^T X \theta.$$

By setting $\frac{\partial L(\theta|X,Y)}{\partial \theta}$ to zero, we have a linear equation system for $\hat{\theta}$, i.e.,

$$-2X^T Y + 2X^T X \hat{\theta} = 0. \tag{3.6}$$

By assumption that $X^T X$ is invertible, the above equation implies that

$$\hat{\theta} = (X^T X)^{-1} X^T Y. \qquad \square$$

Now we understand how to obtain the optimal parameter $\hat{\theta}$, and in practice we simply directly use Equation 3.4 to calculate the optimal parameters from the training data $\mathcal{D}$. The last step is to choose the indicator of the goodness of fit and to evaluate this metric for both training and test sets. The common choices for goodness of fit in OLS are the root mean squared error (RMSE), R^2 and the adjusted R^2.

In summary, the key elements of OLS is provided in Table 3.1.

3.1.2 Pros and cons

In this subsection, let us discuss the pros and cons of OLS. OLS has the advantage of the simplicity of the model and the analytic formula of the optimal parameters, which makes the computation for optimal parameters relatively straightforward compared with other regression methods. Besides, due to the linear model of OLS, it has great interpretability, and one may

Table 3.1. Ordinary Least Squares (OLS).

Dataset:	$\mathcal{D} = \{(x_i, y_i)\}_{i=1}^N$.
Model:	$y = f_\theta(x) + \varepsilon = x\theta + \varepsilon$.
Loss Function:	$L(\theta|X, Y) = (Y - X\theta)^T(Y - \theta) \to \min$.
Optimization:	$\hat{\theta} = (X^T X)^{-1} X^T y$.
Prediction:	$\hat{y}_* = x_* \hat{\theta}$.
Validation:	Compute RMSE, R^2, adjusted R^2 or p-value.

tell which input variables contribute to impacting the output most significantly simply by ranking the corresponding linear coefficients.

However, OLS also has various limitations. First of all, the linear model may be too simple to model the relationship between the input and the output in real world problems. Thus it has the potential risk of the so-called *underfitting* issue.

Secondly, even if a linear model is a reasonable model, by Lemma 3.1, OLS can't be directly applied when $X^T X$ might not be invertible. Let's investigate the possible reasons for $X^T X$ not being invertible as this will help us understand under which situations OLS is not applicable. There are two sufficient conditions for $X^T X$ not being invertible: first, the linear co-dependence of the input data; second, that the dimension of the input variable d is strictly larger than the sample size N, which makes Equation 3.6 an under-determined linear equation system for $\hat{\theta}$, which in turn may allow infinitely many possible $\hat{\theta}$. In this situation, randomly picking one possible $\hat{\theta}$ may lead to little predictive power of the estimated model in the test dataset (overfitting issue). One possible way to solve the problem of the co-linearity of the input data is to use principal component analysis (PCA) for dimension reduction, see Section 7.2 for more details.

Lastly, the loss function of OLS is defined to be an equally weighted squared residual, which results in it being sensitive to outliers. Possible solutions include preprocessing the data in order to remove outliers or modifying the loss function to take into account the possibility of the occurrence of outliers.

A summary of the advantages and disadvantages of OLS is given in Table 3.2.

3.2   Linear Model with Regularization

In the previous subsection, we discuss the potential problems of OLS. The overfitting problem arises when the dimension of the input is strictly larger than the sample size. Here we explain how to use the regularization method to help with this issue.

3.2.1   Regularization

Regularization is a process of introducing additional constraints of the norm of the parameterized function family in order to solve such kinds of ill-posed problems and prevent overfitting.

Table 3.2. Pros and cons of OLS.

Pros

- The linear model is simple and less likely to have the overfitting issue.
- It gives an analytic formula for the optimal parameters, and the computation cost is cheap.
- It is easy to understand and has great interpretability.

Cons

- The linear model may be too restrictive to describe data (**Underfitting issue**).
- Even if the linear model is a reasonable model, OLS has a problem when $X^T X$ is not invertible:
  - There is co-linear dependence between different coordinates of the input variables (**Dimension Reduction**).
  - The dimension of the input space d is larger than the sample size N (**Overfitting issue**).
- It is sensitive to outliers.

The regularized linear regression method is based on OLS, but it adds additional constraints to the optimization process. Specifically, we consider the constraint optimization problem for the least squared loss function:

$$\min_{\beta}(Y - X\beta)^T(Y - X\beta),$$

subject to $||\beta|| \leq t$.

By the Lagrange multiplier, this is equivalent to the unconstrained optimization problem by adding a penalty term to the loss function, i.e., for given $t > 0$,

$$\min_{\beta,\lambda}(Y - X\beta)^T(Y - X\beta) + \lambda(||\beta|| - t).$$

This motivates us to consider the following unconstrained optimization problem, i.e., for given $\tilde{\lambda}$,

$$\min_{\beta,\tilde{\lambda}}(Y - X\beta)^T(Y - X\beta) + \tilde{\lambda}||\beta||, \tag{3.7}$$

where the exact relationship between t and $\tilde{\lambda}$ is data dependent.

In principle, $\tilde{\lambda}$ can be computed directly by knowing t, X and Y. In practice, choosing the proper t is equivalently difficult to the choice of $\tilde{\lambda}$. Thus the procedure of the regularized linear model is outlined as follows:

(1) Specify the set of possible values for $\tilde{\lambda}$, denoted by $\mathcal{L}$.
(2) For any $\tilde{\lambda} \in \mathcal{L}$, solve the unconstrained optimization problem given in Equation 3.7 and obtain the optimal parameter $\beta(\tilde{\lambda})$ and the corresponding $\mathrm{MSE}(\tilde{\lambda})$.
(3) Apply cross validation to choose the best $\tilde{\lambda}$ according to the corresponding MSE, and the corresponding optimal parameter is set to be $\hat{\beta}$, where

$$\hat{\beta} = \beta(\arg\min_{\tilde{\lambda} \in \mathcal{L}} \mathrm{MSE}(\tilde{\lambda})).$$

The universe of regularized linear regression methods is categorized according to the norm of parameters in the penalty term. The l_p norm is the standard one to be used in this context. Let us give the definition of the l_p norm. The l_p norm of any element $x \in \mathbb{R}^d$, denoted by $||x||_p$, is defined as follows:

$$||x||_p = \left(\sum_{i=1}^{d} |x^{(i)}|^p \right)^{1/p}.$$

The three major types of regularized linear regression are listed in Table 3.3. We focus on the first two methods, i.e., Ridge Regression and Lasso Regression, compare their strengths and weakness and establish the link between them based on recent work [Hoff (2017)].

3.2.2 Ridge Regression

Let us consider the Ridge Regression where the loss function is defined as

$$L(\beta|X,Y) = (Y - X\beta)^T(Y - X\beta) + \underbrace{\lambda||\beta||_2^2}_{l_2 \text{ regularization}},$$

where $\lambda > 0$ is a hyperparameter. Thanks to the differentiability of the squared l_2 norm, the parameter estimator of Ridge Regression also yields an analytic closed formula just as that of OLS (Lemma 3.2).

Lemma 3.2 (Ridge Regression Estimator). *Fix $\lambda > 0$. Given the input–output data set $\mathcal{D} = (X, Y)$, and assuming the standard setting of*

Table 3.3. Main types of regularized linear regression methods.

- Ridge Regression:
$$L(\beta|X,Y) = (Y - X\beta)^T(Y - X\beta) + \underbrace{\lambda||\beta||_2^2}_{l_2 \text{ regularization}} ;$$

- Lasso Regression:
$$L(\beta|X,Y) = (Y - X\beta)^T(Y - X\beta) + \underbrace{\lambda||\beta||_1}_{l_1 \text{ regularization}} ;$$

- Elastic Net:
$$L(\beta|X,Y) = (Y - X\beta)^T(Y - X\beta) + \lambda\underbrace{\left(\frac{1-\alpha}{2}||\beta||_2^2 + \alpha||\beta||_1\right)}_{l_1 \text{ and } l_2 \text{ regularization}}.$$

linear regression model holds, the estimator of the optimal parameter $\hat{\theta}$ in Ridge Regression is given as follows:

$$\hat{\theta} = (X^TX + \lambda I)^{-1}X^TY, \tag{3.8}$$

where I is an identity matrix.

Proof. Expand the loss function $L(\beta|X,Y)$ and it is easy to see that it is a quadratic function with respect to β:

$$L(\beta|X,Y) = (Y - X\beta)^T(Y - X\beta) + \lambda||\beta||_2^2$$
$$= Y^TY - 2\beta^TX^TY + \beta^TX^TX\beta + \lambda\beta^T\beta,$$

where $\lambda > 0$ is a constant.

Similar to the proof of the OLS estimator of Lemma 3.1, the optimal parameter $\hat{\beta}$ should satisfy that the first derivative of $L(\beta|X,Y)$ with respect to β, denoted by $\nabla L(\beta|X,Y)$, evaluated at $\beta = \hat{\beta}$ is equal to 0.

Take the first derivative of $L(\beta|X,Y)$ with respect to β and we have

$$\nabla L(\beta|X,Y) = -2X^TY + 2X^TX\beta + 2\lambda\beta$$
$$= 2\left(-X^TY + (X^TX + \lambda I)\beta\right),$$

where I is an identity matrix.

Here for $\lambda > 0$, $X^TX + \lambda I$ is invertible. Setting $\nabla L(\beta|X,Y) = 0$, we have the optimal β^* given as follows:

$$\beta^* = (X^TX + \lambda I)^{-1}X^TY. \qquad \square$$

Table 3.4. Ridge Regression.

Dataset:	$\mathcal{D} = \{(x_i, y_i)\}_{i=1}^{N}$.
Model:	$y = f_\theta(x) + \varepsilon = x\theta + \varepsilon$.
Loss Function:	$L(\theta|X,Y) = (Y - X\theta)^T(Y - X\theta) + \lambda||\theta||_2^2 \to \min$.
Optimization:	$\hat{\theta} = (X^TX + \lambda I)^{-1}X^Ty$.
Prediction:	$\hat{y}_* = x_*\hat{\theta}$.
Validation:	Compute RMSE, R^2, the adjusted R^2 or p-value.

Remark 3.2. When $\lambda = 0$, Ridge Regression is reduced to OLS. Even if X^TX is not invertible, for $\lambda > 0$, $X^TX + \lambda I$ is invertible, which ensures validity of Ridge Regression and solves the potential overfitting problem of OLS.

A summary of Ridge Regression is provided in Table 3.4.

3.2.3   Lasso Regression

The loss function of Lasso Regression is defined as

$$L(\beta|X,Y) = (Y - X\beta)^T(Y - X\beta) + \underbrace{\lambda||\beta||_1}_{l_1\,\text{regularization}},$$

where $\lambda > 0$ is a hyperparameter. We can see that the only difference between these two methods is the form of parameters in the penalty term.

In general, Lasso Regression does not have a closed form solution so that numerical techniques are required. In contrast to Lasso Regression, Ridge Regression admits an analytic solution for the optimal parameters (Lemma 3.2). The benefit of Lasso is that some of coefficients are pushed to exactly zero. In this way Lasso can be used for feature selection. The reason Ridge Regression solutions do not hit exactly zero but Lasso solutions do is due to the different natures of the geometries of the l_1 norm and l_2 norm (see Figure 3.1).

In the two-dimensional case, the difference can be viewed in the figure below. Lasso Regression constraints make a diamond aligned with the axis (Figure 3.1a). The contours inscribed by the solutions (the blue circles in Figure 3.1b) can easily intersect the diamond at its corner. This implies that the Lasso solution forces some coordinates of optimal linear coefficients to zero. This is much more unlikely in the case on the right for Ridge Regression where the l_2-norm constraint inscribes a circle, which is shown in Figure 3.2.

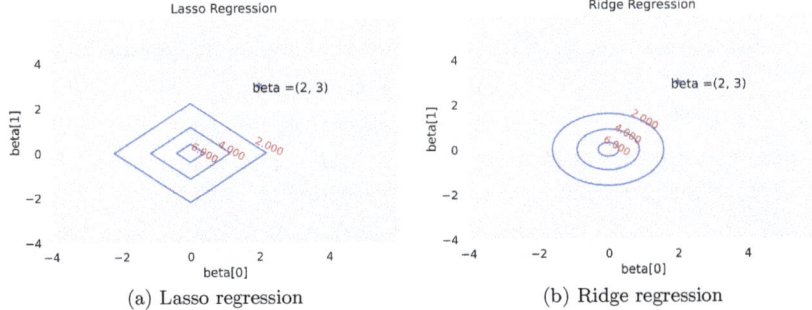

Figure 3.1. Geometry of the Lasso and Ridge regression solutions.[1]

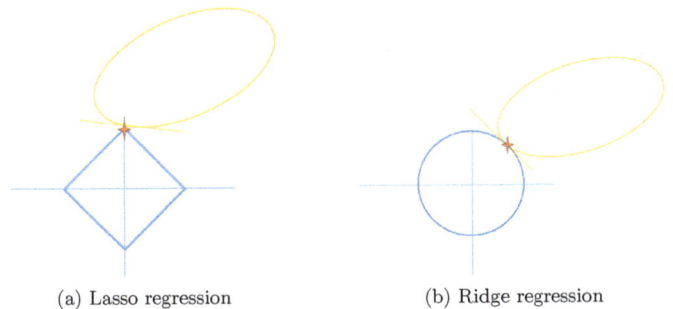

(a) Lasso regression   (b) Ridge regression

Figure 3.2. Typical cases of the Lasso and Ridge regression solutions.

Table 3.5. Comparison of the three regularized linear regression methods.

- Lasso Regression requires numerical techniques to solve, but it can be used for feature selection.
- Ridge Regression has a closed-form solution for the optimal parameters.
- Elastic Nets is a combination of Lasso and Ridge regression.

The comparison of different regularization methods is summarized in Table 3.5.

3.2.4 Numerical example

In this section, we use synthetic data to simulate a sparse linear model, where most of the linear coefficients of this model are zero. We can see how the different regularized linear regression methods perform compared with OLS in this example.

---

[1]https://jamesmccammon.com/2014/04/20/lasso-and-ridge-regression-in-r/.

Example 3.1 (Sparse Linear Model). We simulate 1200 samples of the input–output pairs $(x_i, y_i)_{i=1}^{1200}$ based on the following sparse linear model:

$$y = x\beta + \varepsilon,$$

where $x \in \mathbb{R}^{800}, y \in \mathbb{R}$ and β is a sparse vector, i.e., there are only three non-zero coordinates of β. Here $[\beta^{(1)}, \beta^{(5)}, \beta^{(10)}] = [1.0, 3.0, 2.7]$. We use 80% of data for the training set and the rest for the test set.

We use the Scikit-Learn python package to implement the above linear regression conveniently. The key part of the python code is given in Listing 3.2.

```
1   # Skip the code for simulation of data # Split the dataset into the
    ↪   training set
2   and test set from sklearn.model_selection import train_test_split X_train,
    ↪   X_test,
3   Y_train, Y_test= train_test_split(X, Y, test_size = 0.2)
4
5   ## OLS Regression from sklearn.metrics import r2_score, mean_squared_error
    ↪   from
6   sklearn import linear_model model =
    ↪   sklearn.linear_model.LinearRegression()
7   model.fit(X_train, Y_train) Y_test_pred = model.predict(X_test) print("r^2
    ↪   on test
8   data : {}".format(r2_score(Y_test,Y_test_pred)) print("rmse on test data :
9   {}".format( np.sqrt(mean_squared_error(Y_test, Y_test_pred))))
10
11  ## Ridge Regression from sklearn.linear_model import RidgeCV ridgecv =
12  RidgeCV().fit(X_train, Y_train) Y_test_pred_Ridge =
    ↪   ridgecv.predict(X_test)
13  print("r^2 on test data : {}".format( r2_score(Y_test,
    ↪   Y_test_pred_Ridge)))
14  print("rmse on test data : {}".format( np.sqrt(mean_squared_error(Y_test,
15  Y_test_pred_Ridge))))
16
17  ## Lasso Regression from sklearn.linear_model import Lasso, LassoCV from
18  sklearn.feature_selection import SelectFromModel lassocv =
19  LassoCV(cv=20).fit(X_train, Y_train) Y_test_pred_Lasso =
    ↪   lassocv.predict(X_test)
20  print("r^2 on test data : {}".format( r2_score(Y_test,
    ↪   Y_test_pred_Lasso)))
21  print("rmse on test data : {}".format( np.sqrt(mean_squared_error(Y_test,
22  Y_test_pred_Lasso))))
```

Listing 3.2. Python code for OLS, Ridge Regression and Lasso Regression using the Scikit-Learn package.

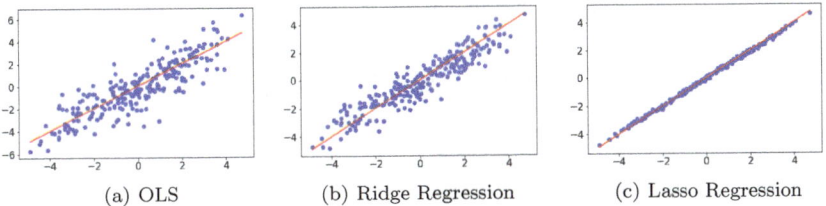

(a) OLS　　　　　　(b) Ridge Regression　　　　　　(c) Lasso Regression

Figure 3.3. Comparison of OLS, Lasso Regression and Ridge Regression in Example 3.1. x-axis and y-axis represent the actual mean function of the output and the model estimated output, respectively. The red line is the identity function.

Table 3.6. Summary of the R^2 and RMSE of the three linear regression methods.

	R^2	RMSE
OLS	0.6561	1.2614
Ridge	0.8197	0.9132
Lasso	0.9461	0.4993

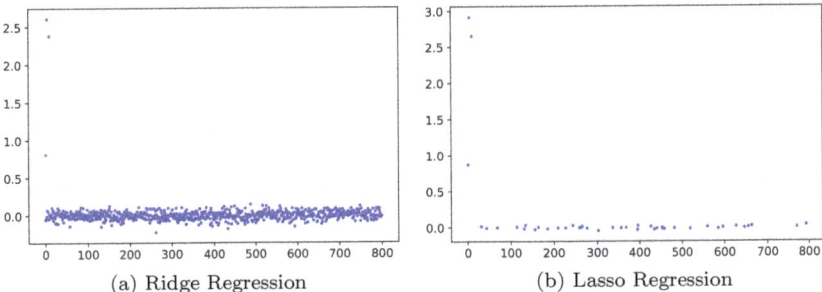

(a) Ridge Regression　　　　　　　　　(b) Lasso Regression

Figure 3.4. Visualization of linear coefficients in regularized regression methods.

In Figure 3.3, we plot the estimated output against the actual mean function of the output using OLS, Lasso Regression and Ridge Regression on the test dataset. The red line in each subplot represents the identity function, which corresponds to a perfect fit. The closer the blue cloud of points is to the red line, the better the corresponding regression method. Figure 3.3 suggests that Lasso works best in this example, as does the summary of the R^2 and RMSE of the three methods (Table 3.6).

Figure 3.4 plots the non-zero linear coefficient estimator across the corresponding index of the input variable coordinate. It supports one of the main differences between Lasso and Ridge Regression, i.e., the linear coefficient estimators of Ridge Regressions are dense, and many of them are close to zero but not exactly zero, while the Lasso forces many of the linear coefficient estimators to be exactly zero. In the situation similar to Example 3.1 where the data is simulated based on the sparse linear model, Lasso Regression is preferable to Ridge Regression due to its capacity for feature selection.

3.2.5 Connection between Ridge Regression and Lasso Regression

In the last part of this subsection, we discuss a new algorithm to compute the Lasso Regression estimator, which explores an elegant link between Ridge Regression and Lasso Regression. Following [Hoff (2017)], Lasso Regression can be formulated as alternating Ridge Regressions. It relies on a the key observation that bridges the loss function of both Ridge Regression and Lasso Regression (Lemma 3.3). Let $Q(\beta)$ denote the mean squared error, i.e.,

$$Q(\beta) := Q(\beta|\mathcal{D}) = (Y - X\beta)^T(Y - X\beta).$$

Lemma 3.3.

$$\min_{\beta} \{Q(\beta) + \lambda|\beta|_1\} = \min_{u,v} \left\{ Q(u \circ v) + \frac{\lambda}{2} \left(|u|_2^2 + |v|_2^2 \right) \right\}, \text{ and } \beta^* = u^* \circ v^*,$$

where $\circ$ is the Hadamard (elementwise) product of the vectors u and v; β^*, u^* and v^* are optimal parameters.

For simplicity we prove Lemma 3.3 for the one dimensional output case, which can easily be generalized to the multi-dimensional case. The interested reader may find additional relevant information on the following website.

For scalar β, Lemma 3.3 is simplified to

$$\min_{u,v} \left\{ Q(uv) + \frac{\lambda}{2} \left(u^2 + v^2 \right) \right\} = \min_{\beta} \{Q(\beta) + \lambda|\beta|\},$$

where $Q(\beta) = X^T X \beta^2 - 2X^T Y \beta + Y^T Y$.

Proof. The proof is divided into two parts. Let us first show that

$$\min_{u,v} Q(uv) + \frac{\lambda}{2} \left(u^2 + v^2 \right) \geq \min_{\beta} Q(\beta) + \lambda|\beta|.$$

This follows via

$$\min_{u,v} Q(uv) + \frac{\lambda}{2}\left(u^2 + v^2\right) = \min_{\beta=uv} Q(\beta) + \lambda\left(u^2 + \frac{\beta^2}{u^2}\right)$$

$$\geq \min_{\beta} Q(\beta) + \lambda\sqrt{\beta^2} = \min_{\beta} Q(\beta) + \lambda|\beta|,$$

where the minimum is achieved when $u^2 = \beta = v^2$.

Next let us show the opposite direction of the equality. For any given β, choose $u = \sqrt{|\beta|}$ and $v = \frac{\beta}{u}$; then

$$Q(uv) + \frac{\lambda}{2}\left(u^2 + v^2\right) = Q(\beta) + \lambda|\beta|.$$

Then it follows that

$$\min_{\beta} Q(\beta) + \lambda|\beta| = \min_{u=\sqrt{|\beta|},v=\beta/u} Q(uv) + \frac{\lambda}{2}\left(u^2 + v^2\right).$$

As the constrained minimum is larger than the unconstrained minimum, it holds that

$$\min_{u=\sqrt{|\beta|},v=\beta/u} Q(uv) + \frac{\lambda}{2}\left(u^2 + v^2\right) \geq \min_{u,v} Q(uv) + \frac{\lambda}{2}\left(u^2 + v^2\right), \quad (3.9)$$

which concludes the second part of the argument. □

Theorem 3.1 (Lasso Estimates via Alternating Ridge Regressions). *Let $\hat{\beta}$ denote the Lasso estimator of the linear coefficient, i.e., for $\lambda > 0$*

$$\hat{\beta} := \arg\min_{\beta}\left\{Q(\beta) + \lambda|\beta|\right\}.$$

Then it follows that

$$\hat{\beta} = \hat{u} \circ \hat{v}, \qquad (3.10)$$

where

$$(\hat{u}, \hat{v}) = \arg\min_{u,v}\left\{Q(u \circ v) + \frac{\lambda}{2}\left(|u|_2^2 + |v|_2^2\right)\right\}.$$

For fixed v, the optimal $\tilde{u}$ is given as

$$\tilde{u} = \left(X^T X \circ vv^T + \frac{\lambda I}{2}\right)^{-1}(v \circ X^T Y),$$

where I is the identity matrix.

Proof. Equation (3.10) follows due to Lemma 3.3. For the scalar β case, when v is fixed, the Hadamard product $\circ$ is the scalar multiplication, and

$$\tilde{Q}(u) := \left\{ Q(u \circ v) + \frac{\lambda}{2} \left(|u|_2^2 + |v|_2^2 \right) \right\}$$

$$= (uv)^2 X^T X - 2uv X^T Y + Y^T Y + \frac{\lambda}{2} \left(u^2 + v^2 \right)$$

is a quadratic function with respect to u. The optimal $\tilde{u}$ to minimize $\tilde{Q}(u)$ is $\tilde{u}$ such that

$$\tilde{Q}'(u)|_{u=\tilde{u}} = 0.$$

Therefore it follows that

$$\tilde{Q}'(u)|_{u=\tilde{u}} = \left(2v^2 X^T X + \lambda I \right) u - 2v X^T Y = 0.$$

Rearranging the above equation, we have that

$$\tilde{u} = \left(X^T X \circ vv^T + \frac{\lambda I}{2} \right)^{-1} (v \circ X^T Y). \qquad \square$$

Theorem 3.1 demonstrates that the Lasso solution is equivalent to the minimum of a certain quadratic function where the minimum is taken over all possible (u, v). In addition, according to Theorem 3.1 for a given v, the optimal $\tilde{u}$ can be regarded as the Ridge Regression criterion $(\tilde{Q}(u))$ where u is the parameter. A similar result holds for the optimal value of v given u. An alternate Ridge Regression algorithm for finding the Lasso estimator $\hat{\beta} = \hat{u} \circ \hat{v}$ is therefore to iterate the following algorithm to update u_n and v_n alternately until convergence:

(1) Set $v_n = (X^T X \circ u_n u_n^T + \frac{\lambda}{2} I)^{-1} (X^T Y \circ u_n)$.
(2) Set $u_n = (X^T X \circ v_n v_n^T + \frac{\lambda}{2} I)^{-1} (X^T Y \circ v_n)$.

The corresponding algorithm is summarized to Algorithm 3.

3.3 Extension of Linear Models: Basis Expansion

In the previous subsection, we studied linear regression and regularization methods to avoid potential overfitting issues. However, for some cases, linear regression suffers from a potential *underfitting* problem. Suppose the actual mean function is non-linear; applying linear regression leads to the risk of of model misspecification.

Algorithm 3: Input(X, Y, λ, N_e)

1: Initialize u_0;
2: **for** $n = 1 : N_e$ **do**
3:     Compute v_n and u_{n+1} based on the following equations

$$v_n = \left(X^T X \circ u_n u_n^T + \frac{\lambda I}{2} \right)^{-1} (u_n \circ X^T Y),$$

$$u_n = \left(X^T X \circ v_n v_n^T + \frac{\lambda I}{2} \right)^{-1} (v_n \circ X^T Y).$$

4: **end for**
5: Set $\hat{\beta} = v_n \circ u_n$.
6: **return** $\hat{\beta}$.

Therefore it is essential to propose good non-linear models to fit the data. The main idea shared by various non-linear models is the basis expansion, which is to linearize the functional relationship between the input and the output by using a non-linear transformation mapping the input to the features. As a formula:

$$y = f(x) + \varepsilon,$$

$$f(x) \approx f_\theta(x) := \sum_{i=1}^{n} \theta_i \phi_i(x), x \in \mathbb{R}^d,$$

where $\phi: x \mapsto (\phi_1(x), \ldots, \phi_n(x))$ is a feature map, and $(\theta_i)_{i=1}^{n}$ are model parameters. There are two types of basis function, i.e.,

- Fixed basis functions (features), e.g.,
  - Polynomial basis: $x^0, x^1, x^2, \ldots, x^n$;
  - Spline basis...

- Adaptive basis functions, e.g., Neural Networks. Figure 2.2 describes different types of Neural Networks.

3.4 Exercises

(1) Explain the ordinary least square method (OLS).
(2) Explain at least two ways to relieve the overfitting issues for linear regression methods.
(3) Implement the Elastic-Net Regression and apply it to Example 3.1.

(4) Prove Theorem 3.1 for $\beta \in \mathbb{R}^d$.

(5) Name three non-linear regression methods that are not covered in this chapter and summarize their model, loss function, optimization and goodness of fit metric.

Chapter 4

Tree-based Models

4.1 Introduction

Tree-based methods, which utilize a tree structure to build a predictive model, are popular for supervised learning problems. A key advantage of the tree-based model is its interpretability. The simplest tree-based method is the decision tree model, and more advanced tree-based methods include random forest, gradient boosting trees and others. The decision tree model is straightforward to understand even for people from non-analytical backgrounds, and mimics the way a clinician uses a flowchart to access the health status of patients. It does not require any statistical knowledge to interpret them, and their graphical representation is very intuitive. Thus the decision tree is called a white-box model, in contrast to black-box models, e.g., neural networks, which are discussed in the next chapter.

Besides this, the tree-based model provides a natural way to rank all the features based on their importance scores, which is useful for variable selection. Decision trees can help create new variables/features that have better power to predict the target variable. They can also be used in the data exploration stage—for example, when we are working on a problem where there are hundreds of variables in the dataset. In such a case, tree-based models can help to select the most significant variables.

Furthermore, it requires less data cleaning compared to some other modeling techniques. It is robust to outliers and missing values to a fair degree. Moreover, it is non-parametric, and it can handle both numerical and categorical variables so is applicable to both regression and classification problems.

Table 4.1. Pros and cons of tree-based models.

Pros
• Easy to understand (interpretability).
• Useful in data exploration and good for feature selection.
• Less data cleaning required.
• Data type is not a constraint.
• Non-parametric method.
Cons
• Over-fitting.
• Not suitable for continuous variables.

Although the tree-based model has various strengths, it suffers from the following potential drawbacks. First of all, it may lead to the overfitting issue. This problem may be alleviated by setting constraints on model parameters and 'pruning'. Besides, due to the discrete nature of the decision tree, for continuous numerical variables, decision trees lose information when they use discrete variables to approximate a continuous input/output. A summary of the tree-based method and its advantages and disadvantages can be found in Table 4.1.

In the following, we first introduce the decision tree method for supervised learning. Then we discuss the ensemble methods using the decision tree as the base learner—the *random forest* and the *gradient boosting decision tree*.

4.2 Decision Tree

The tree is a commonly used data structure, which is defined recursively. In this section, we first introduce the definition of a tree and the essential terminologies of the tree structure in Section 4.2.1. In the subsequent subsections, we continue to focus on the simplest tree-based model, the so-called decision tree method. We explain the models, training and evaluation of the decision tree method and end with a summary of its pros and cons.

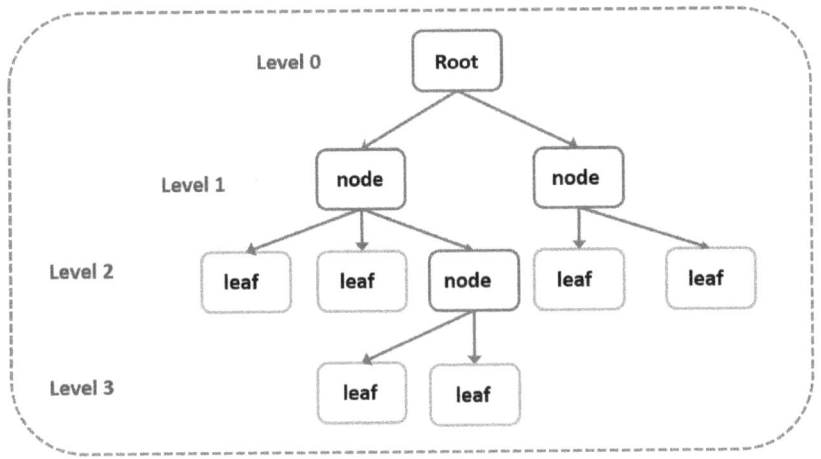

Figure 4.1. Tree structure example.

4.2.1 Tree structure

Let us first introduce the definition of a root tree, whose illustrative example is given in Figure 4.1. A root tree $\mathcal{T} = (V, E)$ is a directed graph, such that for any two nodes there exists only one path connecting them, where V is a set of nodes, and E is a set of directed edges, where each edge e is represented as a pair of nodes (v_1, v_2), with both $v_1, v_2 \in V$.

The alternative definition of a tree is given in a recursive manner:

Definition 4.1 (Tree). A tree of height -1 is an empty set. A tree of height 0 has a unique node (called the root). A tree $\mathcal{T} = (V, E)$ of height h is the composition of a finite number of trees $(\mathcal{T}_i = (V_i, E_i))_{i=1}^{n}$, where $\max_i h_i < h$ and there exists i_* such that $h_{i_*} = h - 1$; there exists a node R_0 (root), such that E is the union of (E_i) and the edges connecting R_0 and the root of $\mathcal{T}_i$, and V is the union of (V_i) and R_0, i.e.,

$$V = \{R_0\} \overset{n}{\underset{i=1}{\cup}} V_i,$$

$$E = \overset{n}{\underset{i=1}{\cup}} \{(R_0, R_i)\} \overset{n}{\underset{i=1}{\cup}} E_i.$$

Remark 4.1. A tree is an undirected graph in which any two vertices are connected by exactly one path.

The basic terminology used with trees is as follows.[1] There are several important types of nodes:

- Root node: This is the node of height 0 in a tree. It represents the entire population or sample and gets further divided into two or more sub-trees.
- Decision node: If a sub-node splits into further sub-nodes, then it is called a decision node.
- Leaf/Terminal node: It refers to a node that does not split.
- Parent and child node: A node that is divided into sub-nodes is called the parent node of the sub-nodes whereas the sub-nodes are the children of the parent node (Figure 4.2).

There are two major operations on trees, i.e., pruning and splitting. They can be seen as opposite processes:

- Pruning: When we remove sub-nodes of a decision node, this process is called pruning.
- Splitting: This is the process of dividing a node into two or more sub-nodes.

Figure 4.2 illustrates the relationship between a parent node (t) and its child nodes (t_L) and (t_R).

4.2.2   Model

A tree model provides a non-parametric way to describe a mapping from an input space $\mathcal{X}$ to an output space $\mathcal{Y}$. Suppose that we consider a d-dimensional input $\bar{x} = (x^1, x^2, \ldots, x^d) \in \mathcal{X} := \mathbb{R}^d$, where x^i is called

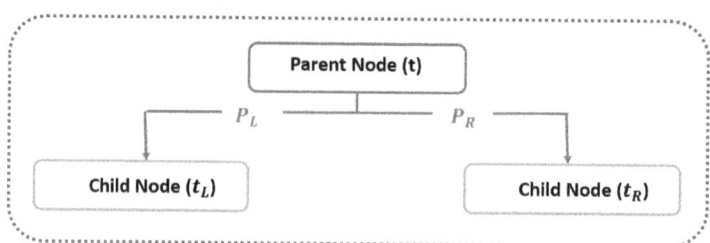

Figure 4.2. Tree unit: Parent node and child nodes.

---

[1]https://www.analyticsvidhya.com/blog/2016/04/complete-tutorial-tree-based-modeling-scratch-in-python/.

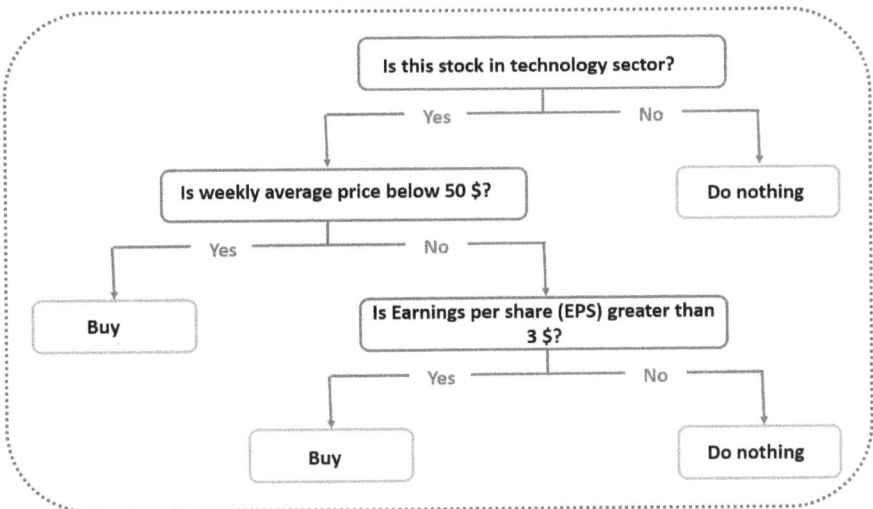

Figure 4.3. Decision tree example of financial applications.

the ith attribute of x. A tree model partitions the input space into a set of rectangles and fits a simple (constant) model in each sub-region of the input space.

A tree model is a flowchart-like structure in which each decision node represents a "test" on an attribute (e.g., whether this stock is in the technology sector in Figure 4.3), each branch represents the outcome of the test and each leaf node represents the output. Each path from the root to a terminal node represents a splitting rule. This model has two important components, a tree $\mathcal{T}$ with M terminal nodes and a parameter set $\Theta = (\theta_1, \theta_2, \ldots, \theta_M)$ that associates the parameter value θ_i with the ith terminal node.

In the following, we focus on the binary tree for concreteness. Following [Chipman *et al.* (1998)], the binary tree $\mathcal{T}$ subdivides the predictor space as follows. Each decision node has an associated splitting rule that uses a predictor to assign observations to either its left or right child nodes. The terminal nodes thus identify a partition of the observation space according to the subdivision defined by the splitting rules. For a continuous input x, the splitting rule is based on a split value s and assigns observations for which $x^i \leq s$ or $x^i > s$ to the left or the right child node. Note that this formulation is general enough to handle arbitrary splitting functions of the form $h(x^i) \leq s$ or $h(x^i) > s$ by simply treating $h(x)$ as another input. For a categorical input, the splitting rule is based on a category subset C, and assigns observations for which $x^i \in C$ and $x^i \notin C$ to the left or the right child node.

4.2.3   Regression decision tree

One main optimization method to determine the optimal decision tree is the *classification and regression tree* (CART). Let us focus on the regression problem, where the output is a continuous variable. The decision tree model for the regression problem is called a *regression tree*. Suppose that we have a decision tree model associated with T with leaves $R_1, R_2, \ldots, R_M$ defined by the following equation:

$$f_\theta = \sum_m c_m \mathbf{I}(X \in R_m),$$

where $(c_m)_{m=1}^M$ are constants.

For each region R_m, the estimated output is a constant c_m, where $m \in \{1, 2, \ldots, M\}$. The model parameters here include a tree T and $\{c_m\}_{m=1}^M$. As we explained in Section 2.1.2, in regression problems, the most common loss function $L(\theta|(X, Y))$ is the mean squared error, that is

$$L(\theta|(X, Y)) = \sum_{i=1}^N (Y_i - f_\theta(X_i))^2.$$

Once $(R_m)_{m=1}^M$ is given, the best estimator c_m to minimize the loss function can be quickly determined. The above equation is rewritten as follows by grouping each input sample based on the region it belongs to,

$$L(\theta|(X, Y)) = \sum_{m=1}^M \sum_{i \in \{1, 2, \cdots, N\} \in I_m} (Y_i - c_m)^2, \tag{4.1}$$

where $I_m = \{i | X_i \in R_m\}$ is the set of indexes such that the corresponding input samples belong to the region R_m, and $\{R_m\}_{m=1}^M$ are disjoint regions.

Minimizing Equation 4.1 is equivalent to minimizing

$$\sum_{i \in \{1, 2, \cdots, N\} \in I_m} (Y_i - c_m)^2,$$

for any $m \in \{1, 2, \ldots, M\}$.

Then the best estimator for c_m, denoted by $\hat{c}_m$, is given by the following equation:

$$\hat{c}_m = \frac{\sum\limits_{x_i \in R_m} y_i}{\sum\limits_{i=1}^{N} \mathbf{1}(x_i \in R_m)} = \mathrm{avg}(y_i|R_m).$$

As we know how to derive $\hat{c}_m$ for given R_m, the decision tree model estimation boils down to the question of finding the optimal partition for a tree. Unfortunately it is not feasible to find a global optimal partition, even for a binary tree. Hence we apply a greedy search to solve this problem.[2]

For simplicity, we only consider a binary tree in the following discussion, but the proposed algorithm can be generalized to trees with more than 2 nodes. The core idea is the following: first, the algorithm splits the training set into two subsets using a single feature $\mathcal{I}_k$ and a threshold t_k. It searches for the pair $(\mathcal{I}_k, t_k)$ to minimize the loss function. We apply the greedy search for all possible attributes j outlined in Algorithm 4, and recursively repeat the following steps to split each terminal code of the tree into a pair of child nodes, until the minimum node size n_{min} is reached.

Algorithm 4: Decision Tree Algorithm

1: **Input:** H, (x, y) and n_{min}.

2: $T = c_0 = \arg\min_c \sum_{i=1}^{N} Q(y_i - c)$.

3: **for** $h = 0 : H - 1$ **do**

4:     Let $R_1, \cdots R_{n_h}$ denote leaves of the tree T.

5:     **for** $i = 1 : n_h$ **do**

6:         $(\mathcal{I}_i, t_i)$ = Optimal Binary Splitting Algorithm$((x_j, y_j)_{j \in R_i})$.

7:         Grow a tree T by splitting a leaf R_i by $(\mathcal{I}_i, t_i)$.

8:         Compute $\hat{c}_i$ for each leaf R_i.

9:     **end for**

10:     **if** the minimum node size n_{min} is reached **then**

11:         **break**

12:     **end if**

13: **end for**

14: **return** T.

---

[2]A greedy algorithm is an algorithmic method for solving an optimization problem in a heuristic way by making the locally optimal choice at each step with the hope that it is close to the optimum.

Slightly differently from other supervised learning methods like linear regression or neural networks, the loss function of the decision tree model is used to quantify the performance of growing a tree iteratively by splitting an existing node into its child nodes.

4.2.3.1 *Classification tree*

The decision tree for a classification problem is called a **classification tree**. The main difference between the regression tree and classification tree is the loss function, which measures the performance of growing a tree by splitting an existing node to its child nodes, and is called an impurity measure. In a classification tree, there are several commonly used forms of impurity measures, e.g., misclassification error, Gini index and cross entropy. Before giving the definition of the above impurity measures, we first introduce the notation $\hat{p}_{mk}$ for the percentage of samples being estimated as class k on region R_m. $\hat{p}_{mk}$ is defined as follows:

$$\hat{p}_{mk} = \frac{1}{N_m} \sum_{x_i \in R_m} \mathbf{I}(y_i \in k),$$

where N_m denotes the number of samples that belong to region R_m.

- Misclassification error:

$$L(\theta|(X,Y)) = \sum_{m=1}^{M} \frac{1}{N_m} \sum_{i \in I_m} \mathbf{1}(y_i \neq c_m) = \sum_{m=1}^{M} (1 - \hat{p}_{mm});$$

- Gini index:

$$L(\theta|(X,Y)) = \sum_{k=1}^{M} \hat{p}_{mk}(1 - \hat{p}_{mk}) = \sum_{k=1}^{M} 1 - \hat{p}_{mk}^2;$$

- Cross entropy:

$$L(\theta|(X,Y)) = -\sum_{k=1}^{M} \hat{p}_{mk} \log(\hat{p}_{mk}).$$

Let us compare the above three types of impurity measures. Consider the following simple case, where all of the samples are from one class. However, there are two estimated classes, and the estimated percentage of samples belonging to the second class is p. The above three measurements for node impurity are $\min(1 - p, p)$, $2p(1 - p)$ and $-p\log(p) - (1 - p)\log(1 - p)$, respectively, which is depicted in Figure 4.4. They are similar, but note that

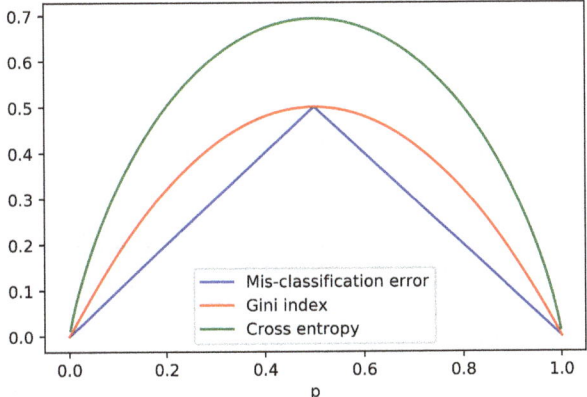

Figure 4.4. Different forms of impurity measures in the classification tree.

unlike misclassification error, Gini index and cross entropy are both differentiable, which is useful for numerical optimization. Besides, the Gini index and cross entropy are more sensitive to changes in the node probability than misclassification error.

Remark 4.2. The Gini index can be interpreted in two different ways. If we classify the observations into class k with probability $\hat{p}_{mk}$, then the training error rate of this strategy is

$$\sum_{k \neq k'} \hat{p}_{mk} \hat{p}_{mk'}, \tag{4.2}$$

which is the Gini index.

Alternatively, if we code each observation as 1 for class k and zeros otherwise, the variance over the node of this indicator variable is $\hat{p}_{mk}(1 - \hat{p}_{mk})$. Summing this up over all possible classes k gives the Gini index.

Having introduced the impurity measures, we are ready to introduce the loss function for the tree classifier. Given any leaf, for any j^{th} attribute and the threshold t_j, we can divide the corresponding samples to the left subset and right subset. The loss function on (j, t_j) is defined as the weighted impurity measure of both left and right subsets as follows:

$$L(j, t_j) = \frac{m_l}{m} G_l + \frac{m_r}{m} G_r, \tag{4.3}$$

where $m = m_l + m_k$

$$\begin{cases} G_{l/r} \text{ measures the impurity of the left/right subset,} \\ m_{l/r} \text{ is the number of the instances in the left/right subset.} \end{cases}$$

For the calibration of a tree classifier, we also use the decision tree algorithm in Algorithm 4, but the optimal binary splitting algorithm in Algorithm 5 for a regression tree needs to be modified to adapt it to the classification case. The only differences are how to compute $c^*_{l/r,j}$ and the loss function $l_{j,n}$. Specifically, line 7 in Algorithm 5 is changed as follows:

Algorithm 5: Optimal Binary Splitting (Regression) Algorithm

1: **Input:** $(x_i, y_i)_{i=1}^N$.

2: Set d to be the dimension of x_1.

3: **for** $j = 1 : d$ **do**

4:      Sort $(x_i, y_i)_{i=1}^N$ based on the jth attribute.

5:      **for** $n = 1 : N$ **do**

6:          Select $x_n^{(j)}$ as the split variable.

7:          Compute c^*_l and c^*_r via

$$c^*_{l,j} = \frac{1}{n} \sum_{i=1}^n y_i,$$

$$c^*_{r,j} = \frac{1}{N-n} \sum_{i=n+1}^N y_i.$$

8:          Compute the loss $l_{j,n}$ by splitting the jth attribute at x_n using $c^*_{l,j}$ and $c^*_{r,j}$ for each sub-region:

$$l_{j,n} = \sum_{i=1}^n Q(y_i - c^*_{l,j}) + \sum_{i=n+1}^N Q(y_i - c^*_{r,j}).$$

9:      **end for**

10:    Set (n_j, s_j) via

$$n_j = \arg\min_n l_{j,n};$$

$$s_j = x_{n_j}.$$

11: **end for**

12: Set j^* to be

$$j^* = \arg\min_n l_{j,n_j}.$$

13: **return** (j^*, s_{j^*}).

Compute c_l^* and c_r^* via

$$c_{l,j}^* \leftarrow \text{the class number which has the most samples in } R_l,$$
$$c_{r,j}^* \leftarrow \text{the class number which has the most samples in } R_r,$$

where region $R_l = \{x_i\}_{i=1}^n$ and region $R_r = \{x_i\}_{i=n+1}^N$.

Line 8 in Algorithm 5 is changed to compute the loss function using Equation (4.3).

4.2.4 Pruning

Tree size is a hyperparameter representing the model complexity. To avoid overfitting, we incorporate the tree size in the loss function, similar to regularized linear regression. The loss function with penalty term in the decision tree model is often given in the following form:

$$L_\alpha(T) = L(T) + \alpha|T|,$$

where $|T|$ denotes the number of leaves in the tree T and $L(T)$ is a loss function without regularization term—e.g., mean squared error in regression.

4.2.5 Feature importance

In the decision tree model, feature importance provides a score for each feature of the data. The higher the score, the more important or relevant is the feature towards the output variable. In the decision tree model, feature importance is calculated by the normalized impurity measure of each feature. Specifically, let us consider a tree model T with leaves $R_1, R_2, \cdots, R_M$. For each (non-terminal) decision node (k, t_k) in the tree T, we compute the corresponding impurity measure gain, i.e.,

$$IG_k := G_a - L(\mathcal{I}_k, t_k) = G_a - \left(\frac{m_l}{m} G_l + \frac{m_r}{m} G_r \right),$$

where G_a is the impurity measure of not splitting the decision node $(\mathcal{I}_k, t_k)$.

Thus for each feature, the following score is defined to be the jth feature importance score:

$$\frac{\sum_{k=1}^M \mathbf{1}_{\mathcal{I}_k = j} IG_k}{\sum_{k=1}^M IG_k}.$$

The feature importance is useful for feature section. We can select the important features of the decision tree based on the feature importance in descending order.

4.3   Random Forest

Random forest is an ensemble method for supervised learning by constructing a multitude of decision trees at training time using bagging, which was introduced in Section 2.3. Let us start with the definition of a forest, which intuitively is a collection of trees, and the definition is given as follows:

Definition 4.2 (Forest). A forest is an undirected graph, all of whose connected components are trees. In other words, the graph consists of a disjoint union of trees.

Random forest uses the decision tree as the base model [Friedman *et al.* (2001)]. It generates multiple decision trees randomly, where the randomization takes place in mainly two ways:

- Random sampling of data for bagging samples.
- Random selection of input features for generating individual base decision trees.

The predictive power of individual decision trees and the correlation among base trees are the two main factors that determine the generalization error of random forest models. The main idea of random forest is to reduce the variance of bagging by reducing the correlation between the trees. The random forest is depicted in Figure 4.5.

The random forest algorithm is given in Algorithm 6.

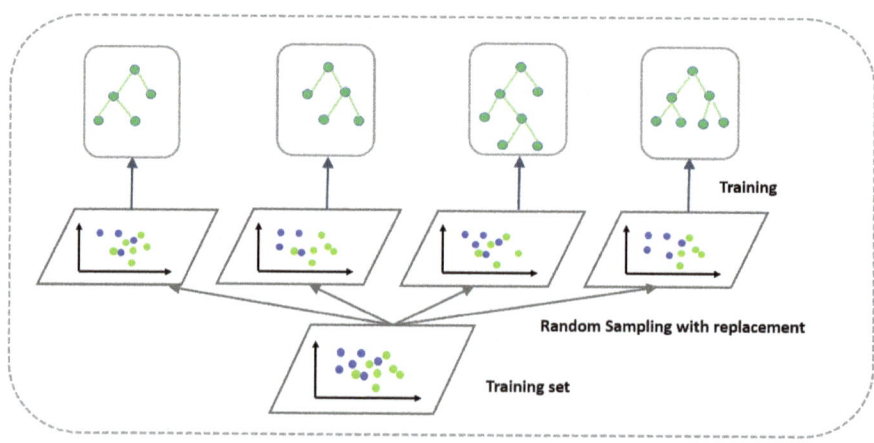

Figure 4.5. Random forest.

Algorithm 6: Random Forest Algorithm

1: **Input:** $H, (x, y), B$ and n_{min}.

2: **for** $b = 1 : B$ **do**

3:     Randomly sample a subset of size N with replacement from the training set, denoted by $\mathcal{D}_b$.

4:     Apply the decision tree algorithm with parameters $(H, \mathcal{D}_b, n_{min})$ to obtain one estimated random-forest tree T_b.

5: **end for**

6: Compute the average model T, i.e.,

$$T = \frac{1}{B} \sum_{b=1}^{B} T_b,$$

or by taking the majority vote in the case of classification trees.

7: **Output:** T.

4.4 Gradient Boosting Decision Tree

In Section 2.3.2, we discussed gradient boosting. The gradient boosting decision tree (GBDT), as its name suggests, is a gradient boosting method using decision trees as base models.

We adopt the notation of the gradient boosting method from Section 2.3. Let f_m denote the estimator for f at the mth iteration. Let r_{im} denote the derivative of $L(y, f(x))$ evaluated at $f = f_m$ for the ith sample, i.e.,

$$r_{im} = \left[\frac{\partial L(y_i, f(x_i))}{\partial f(x_i)} \right]_{f=f_m}.$$

For each leaf of a tree (denoted by R_{jm}), we use a constant model, which can help us to simplify the update rule for the derivative $\frac{\partial L(y_i, f(x_i))}{\partial f(x_i)}$, i.e.,

$$f_m(x) = f_{m-1}(x) + \sum_{j=1}^{J_m} \gamma_{jm} \mathbf{1}(x \in R_{jm}). \tag{4.4}$$

Let h_m denote $\sum_{j=1}^{J_m} \gamma_{jm} \mathbf{1}(x \in R_{jm})$, which is a fitted tree to approximate the residuals at the mth step [Friedman (2001)].

γ_{jm} is taken by

$$\gamma_{jm} = \arg\min_{\gamma} \sum_{x_i \in R_{jm}} L(y_i, f_{m-1}(x_i) + \gamma).$$

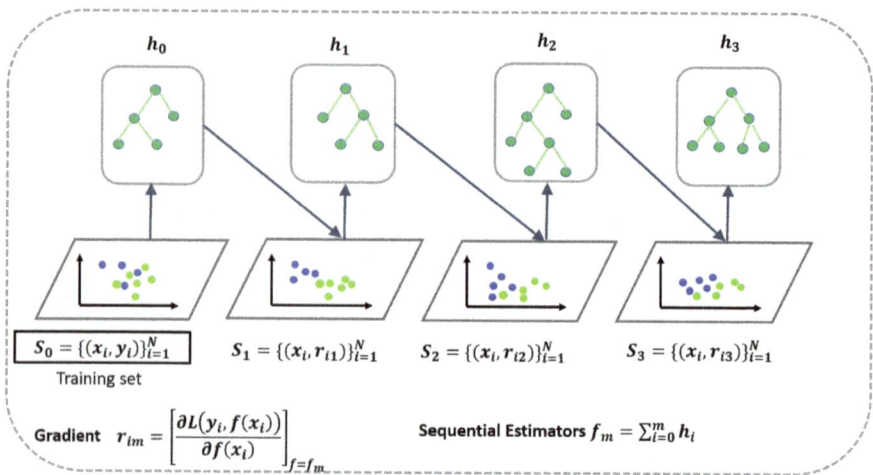

Figure 4.6. Gradient boosting decision tree.

The main idea of GBDT is depicted in Figure 4.6, and GBDT is outlined in Algorithm 7.

There are a few popular and more advanced algorithms based on GBDT, e.g., AdaBoost [Hastie *et al.* (2009)], XGBoost[3] and LightGBM [Ke *et al.* (2017)], which are faster and more efficient. XGBoost and Light GBM have been widely used in machine learning applications and became the most popular algorithms in Kaggle competitions: many Kaggle winners use those two algorithms to win the championships. In Chapter 9, LightGBM will be discussed and applied to solve one Kaggle competition on default risk.

4.5   Numerical Example: Iris Dataset

In this section, we apply the decision tree model, random forest and GBDT to the Iris dataset for classifying flower species,[4] and show how to implement the above tree models for training and prediction. The Iris dataset consists of 150 samples (50 for each species of Iris) and four features (sepal length, sepal width, petal length and petal width).

The financial application of tree models will be introduced in the case study in the last chapter of this book.

---

[3]https://xgboost.readthedocs.io/en/latest/.
[4]https://archive.ics.uci.edu/ml/datasets/iris.

Algorithm 7: Gradient Boosting Decision Tree Algorithm

1: **Input:** $(x_i, y_i)_{i=1}^{N}$.

2: Initialize f_0 by a constant γ_0 via the following equation:

$$\gamma_0 = \arg\min_{\gamma} L(y - \gamma);$$

3: **for** $m = 1 : M$ **do**

4:     **for** $i = 1 : N$ **do**

5:         Compute the residuals

$$r_{im} = \left[\frac{\partial L(y_i, f(x_i))}{\partial f(x_i)} \right]_{f=f_{m-1}}.$$

6:     **end for**

7:     Fit a tree to the target r_{im} giving terminal regions R_{jm}, $j = 1, 2, \cdots, J_m$.

8:     **for** $j = 1, 2, \cdots, J_m$ **do**

9:         Compute

$$\gamma_{jm} = \arg\min_{\gamma} \sum_{x_i \in R_{jm}} L(y_i, f_{m-1}(x_i) + \gamma).$$

10:    **end for**

11:    Update f_m using the following formula:

$$f_m(x) = f_{m-1}(x) + \sum_{j=1}^{J_m} \gamma_{jm} \mathbf{1}(x \in R_{jm}).$$

12: **end for**

13: **Output:** f_M.

4.5.1 Decision tree implementation

The Scikit-Learn library provides a built-in function DecisionTreeClassifier() for the implementation of a decision tree classifier. The code for implementation of the decision tree classifier is outline in Listing 4.3.

Graphviz[5] is open-source graph visualization software. Graph visualization is a common approach to representing structural information. It is a

---

[5]https://www.graphviz.org/.

```
1   import numpy as np
2   import pandas as pd
3   from sklearn.datasets import load_iris
4   from sklearn.model_selection import cross_val_score
5   from sklearn.tree import DecisionTreeClassifier
6   from sklearn.model_selection import train_test_split
7
8   iris = load_iris()
9   X_train, X_test, y_train, y_test = train_test_split(iris.data,
    ↪  iris.target, test_size=0.33, random_state=42)
10
11  clf = DecisionTreeClassifier(random_state=0)
12  clf = clf.fit( X_train, y_train)
13  y_test_est = clf.predict(X_test)
14
15  # Create confusion matrix
16  pd.crosstab(y_test, y_test_est, rownames=['Actual Species'],
    ↪  colnames=['Predicted Species'])
17
18  # View the importance scores
19  print(clf.feature_importances_)
```

Listing 4.3. Implementation of the decision tree classifier using sklearn package.

```
1   from graphviz import Graph
2   dot_data = tree.export_graphviz(clf, out_file=None,
3                          feature_names=iris.feature_names,
4                          class_names=iris.target_names,
5                          filled=True, rounded=True,
6                          special_characters=True)
7   graph = graphviz.Source(dot_data)
8   graph.render("iris")
```

Listing 4.4. Visualization of the tree model using graphviz.

convenient toolbox for visualizing the decision tree model, and the relevant code is given in Listing 4.4. The estimated tree model is depicted in Figure 4.7.

4.5.2   Random forest implementation

Let us continue our discussion on the Iris data and apply random forest to the Iris data. The Scikit-Learn library has RandomForestClassifier() to implement the random forest algorithm, where we can specify the

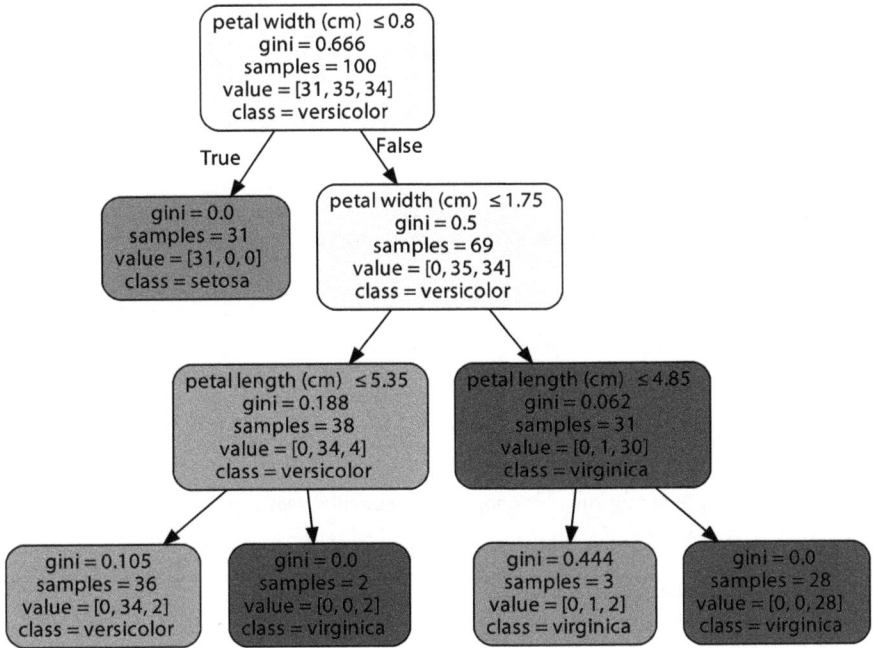

Figure 4.7. The calibrated decision tree model on the Iris dataset.

number of trees (n_estimators), the maximum depth of trees (max_depth) and the impurity measure (criterion). In the Iris data example, we choose n_estimators = 20, max_depth = 3 and criterion = 'gini'. We provide the implementation of the random forest algorithm using Scikit-Learn library in Listing 4.5.

4.5.3 GBDT implementation

We also use the Scikit-Learn built-in function GradientBoostingClassifier() to implement the gradient boosting decision tree for the classification of the Iris data. The code is given in Listing 4.6.

4.5.4 Comparison of three tree-based methods

In this example, the confusion matrices of the testing set for all three methods are the same and are given in Figure 4.8. However, they often have different results in other examples.

```
1    # Load scikit's Random Forest classifier library
2    from sklearn.ensemble import RandomForestClassifier
3
4    # Set random seed
5    np.random.seed(0)
6
7    # Create a Random Forest Classifier.
8    clf_rf = RandomForestClassifier(n_estimators = 20, max_depth = 3,
     ↪   criterion = 'gini', n_jobs=2, random_state=0)
9
10   # Train the Classifier to take the training input (X_train) and learn how
     ↪   they relate
11   # to the training output y_train (the species)
12   clf_rf = clf_rf.fit(X_train, y_train)
13   # Apply the Classifier we trained to the test data (X_test)
14   y_test_est_rf = clf_rf.predict(X_test)
```

Listing 4.5. Implementation of random forest classifier using Scikit-Learn package.

```
1    # Load scikit's tree-based classifier library
2    from sklearn.ensemble import GradientBoostingClassifier
3
4    # Set random seed
5    np.random.seed(0)
6
7    # Create a Random Forest Classifier.
8    clf_gbt = GradientBoostingClassifier()
9
10   # Train the Classifier to take the training input and output (X_train,
     ↪   y_train) and learn how they relate
11   clf_gbt.fit(X_train, y_train)
12   # Apply the Classifier we trained to the test data (X_test)
13   y_test_est_gbt = clf_rf.predict(X_test)
```

Listing 4.6. Implementation of the GBDT classifier using Scikit-Learn package.

Next, let us have a look at feature importance. As we can see from Table 4.2 and Figure 4.9, the petal width is the most important factor for all three tree-based methods. In the decision tree model, sepal length and sepal width are not relevant to forecast the species. Unlike the decision tree, the importance score of sepal length/width is not zero, which is understood as the other two methods are model ensemble methods.

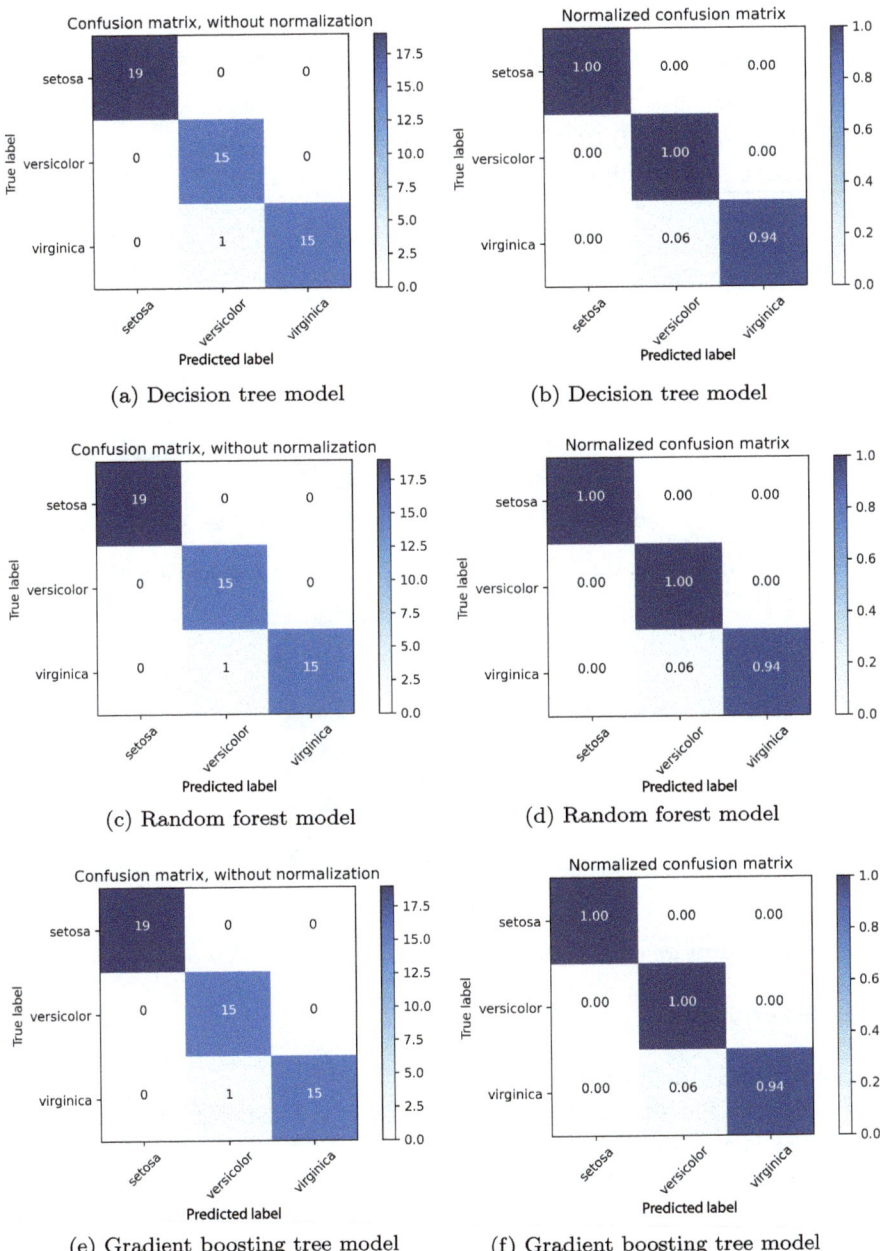

(a) Decision tree model

(b) Decision tree model

(c) Random forest model

(d) Random forest model

(e) Gradient boosting tree model

(f) Gradient boosting tree model

Figure 4.8. Confusion matrix plot for the decision tree, random forest and GBDT on the Iris dataset. The left panels show the unnormalized confusion matrix, while the right panels present the normalized confusion matrix.

Table 4.2. Feature importance for three tree-based models on Iris dataset.

	Sepal length	Sepal width	Petal length	Petal width
Decision tree	0	0	0.0648	0.9352
Random forest	0.1245	0.0308	0.4201	0.4246
GBDT	0.0062	0.0122	0.3046	0.6770

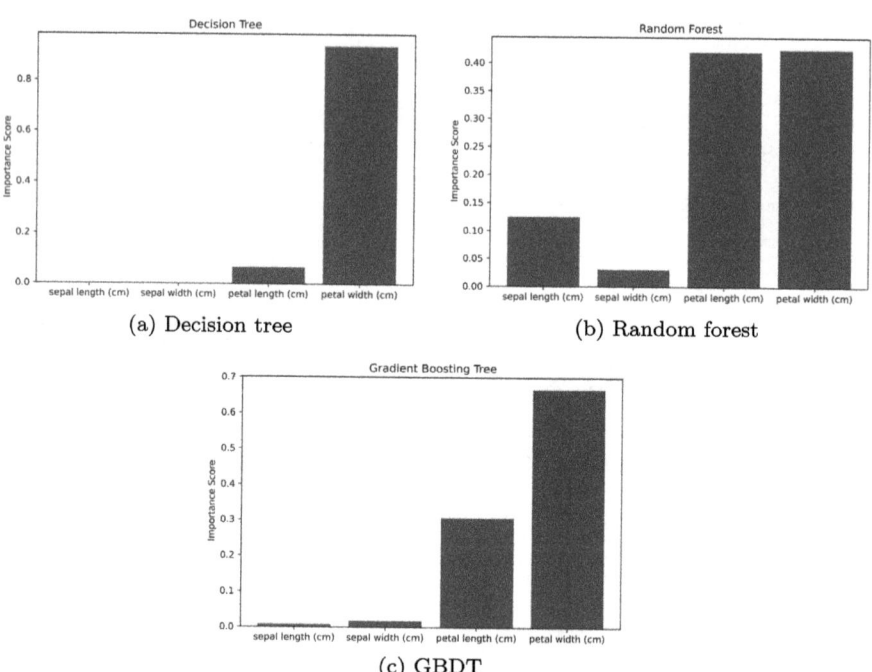

(a) Decision tree

(b) Random forest

(c) GBDT

Figure 4.9. Feature importance plot for three tree-based models on Iris dataset.

4.6 Exercises

(1) What is the definition of a tree?
(2) What are the advantages and disadvantages of the tree-based method?
(3) What is a model ensemble?
(4) What is a random forest?

Chapter 5

Neural Networks

In March 2016, AlphaGo beat Lee Sedol by 4–1 in five games. It is the first time that a computer Go program has defeated a human 9 dan world champion without handicaps. The deep neural network is one of the techniques that AlphaGo used. AlphaGo's victory sparked the discussion on artificial intelligence and made the concept of deep learning well known and accessible to ordinary people all over the world.

Deep learning has a long history, which can be traced back to the 1940s. The development of deep learning is an interesting history, which is full of ups and downs, twists and turns, and successes and failures. There are two peaks while at other times the field remains silent. In the 1940s, the neural network was in its infancy [McCulloch and Pitts (1943)]. Then the perceptron was proposed by [Rosenblatt (1957)], and the network architecture was simply a binary linear classifier (perceptron). In the 1980s, the rise of back-propagation algorithms drove the second wave of neural networks. It seems that now we are climbing the third peak, or perhaps we are already at the summit? For a detailed history of neural networks, interested readers may refer to the following website.[1]

5.1 Basic Terminology

In this section, we provide the basic terminologies of neural networks and defer the rigorous mathematical definition of neural network to the next sections.

---

[1]https://cs.stanford.edu/people/eroberts/courses/soco/projects/neural-networks/History/index.html.

5.1.1 Neuron

A neuron is the basic unit of a neural network. Neural network models are vaguely inspired by biological neural networks that constitute human brains. In the biological neural network, each neuron can detect the environment and exchange signals with other cells through the synapse, where the dendrite receives signals and the axon conveys. Figure 5.1 describes the typical structure of a biological neuron.

The neural network has a similar structure and working mechanism. Each neuron is connected with others, and signals are transferred between them. Let us see the mathematical computation procedure in a neuron. In Figure 5.2, z_1 represents the input (signal) x_1 received by a neuron, and this is transformed to the output h_1 by the activation function σ.

5.1.2 Layer

The multi-layer architecture is one key feature of neural networks compared to most other machine learning algorithms. Several neurons constitute a

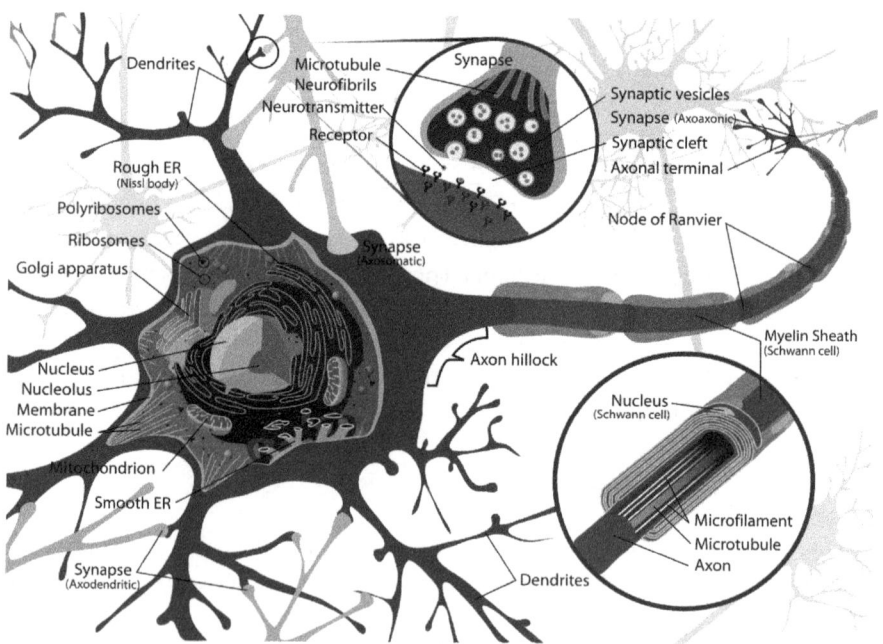

Figure 5.1. Typical structure of a biological neuron.
Source: Retrieved from https://commons.wikimedia.org/wiki/File:Complete_neuron_cell_diagram_en.svg.

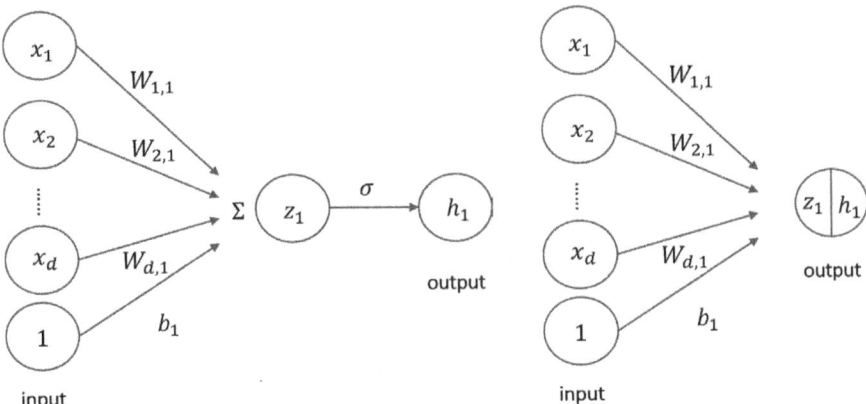

Figure 5.2. An example of the neuron in neuron networks.

layer, and several layers form a multi-layer neural network. The learning process aims to update the optimal parameters of the network, which is a combination of multiple layers. That is why the learning of a neural network is called deep learning.

There are three common types of layers in each neural network, i.e., input layers, hidden layers and output layers. In the convolutional neural network, there are convolutional layers and pooling layers. In the rest of this chapter, we will introduce each of these layers in detail.

Neurons in two consecutive layers may have different kinds of connectivity. If all neurons in the previous layer are connected with those in the next layer, this is called full connectivity. However, if only some of the neurons in one layer are connected with the next layer, this is called local connectivity.

5.1.3 Activation function

The activation function is an essential concept in the neural network. Actually, this concept comes from the perceptron model, which is the basis of the neural network. The perceptron is a binary linear classifier with a concise form. The function f is defined to be $f(x) = \text{sign}(Wx + b)$, i.e., $\forall x \in \mathbb{R}^d$,

$$f(x) = \begin{cases} 1, & Wx + b \geq 0; \\ -1, & Wx + b < 0, \end{cases}$$

where W is a weight parameter and b is a bias term.

Like the sign function in the perceptron model, you can imagine the neuron as an electric circuit and the activation function as a switch. When the neuron receives strong signals, that is $Wx+b \geq 0$, the neuron activates, and the switch opens. In this case, the output of the activation function is 1. When the neuron receives weak signals, it remains inactive, and the output is -1. We can see that the name "activation function" is quite straightforward.

According to the different forms of the activation function, the output of a neuron is not limited to 1 or -1. But the intuition of the activation function is quite similar. The activation functions aim to transform the input by a non-linear map in the elementwise sense. Now we introduce some common activation functions.

5.1.3.1 *Step function*

The Heaviside step function,[2] also called the unit step function, is quite similar to the perceptron and has a simple form, $\forall x \in \mathbb{R}$,

$$f(x) = \begin{cases} 1, & x > 0; \\ 0, & x = 0; \\ -1, & x < 0. \end{cases}$$

One of biggest drawbacks of the Heaviside step function is that $f(x)$ is not differentiable at $x = 0$, and left/right derivatives explode, which limits the use of the Heaviside step function in neural networks.

5.1.3.2 *Sigmoid function*

The sigmoid function, a function having a characteristic S-shaped curve, is an approximation of the step function. It has the following form:

$$\sigma(x) = \frac{1}{1 + e^{-x}}, \forall x \in \mathbb{R}.$$

Figure 5.3a shows the sigmoid function and its derivative. Compared with the Heaviside step function, the sigmoid function has good mathematical properties as it is differentiable everywhere.

---

[2]The Heaviside step function is a special case of the step function. Try not to confuse it with the step function in Theorem 5.1.

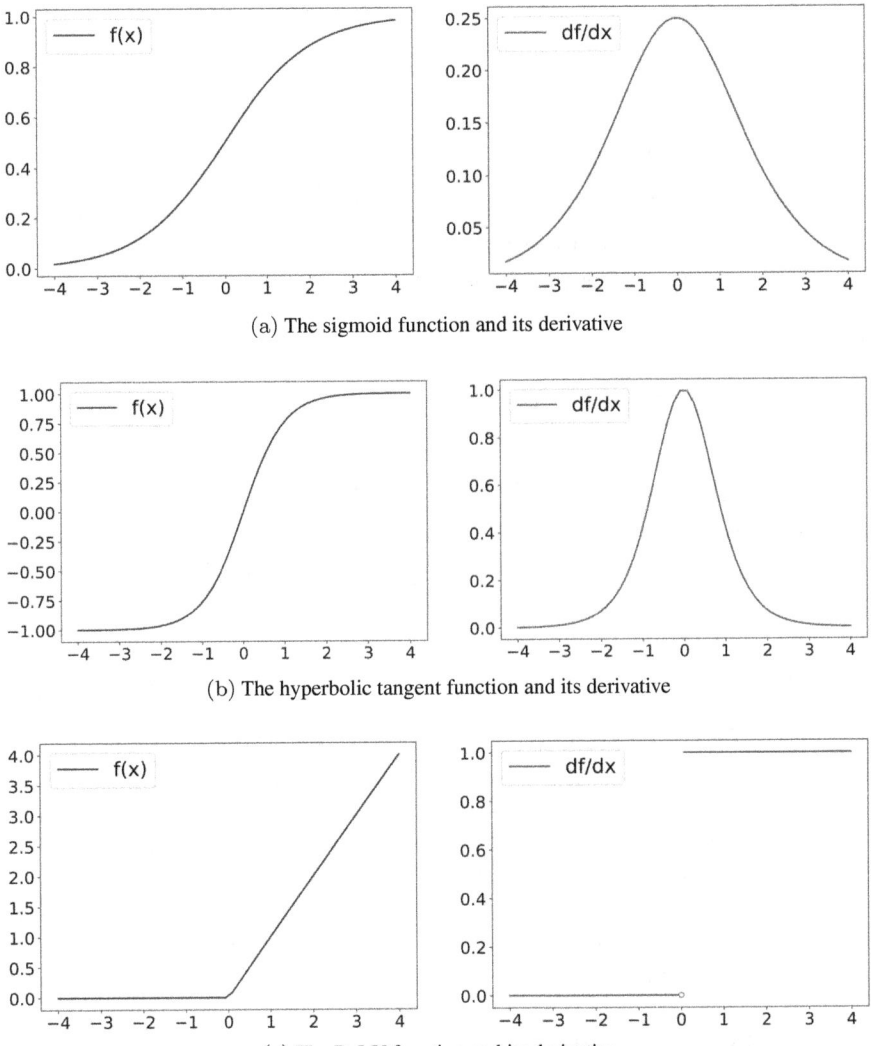

(a) The sigmoid function and its derivative

(b) The hyperbolic tangent function and its derivative

(c) The ReLU function and its derivative

Figure 5.3. Different kinds of activation functions.

However, when the absolute value of the input becomes large, the gradient of the sigmoid function tends to zero. Therefore the learning process of the model is relatively slow, and parameters can hardly be updated, which is called the vanishing gradient problem. It is a common problem in the training of deep neural networks.

5.1.3.3   *Tanh*

The hyperbolic tangent function is a generalization of the sigmoid function and has a similar curve:

$$\tanh(x) = \frac{\sinh(x)}{\cosh(x)} = \frac{e^x - e^{-x}}{e^x + e^{-x}}, \forall x \in \mathbb{R}.$$

As shown in Figure 5.3b, the output of the hyperbolic tangent function is between -1 and 1, and the function is centrosymmetric. But there still exists a vanishing gradient problem.

5.1.3.4   *ReLU*

The ReLU (Rectified Linear Unit) function is one of the most common activation functions, and has the following form:

$$f(x) = \max(x, 0), \forall x \in \mathbb{R}.$$

From Figure 5.3c, we can see that the ReLU function solves the vanishing gradient problem and is often used in deep neural networks.

5.1.3.5   *Softmax*

The softmax function, also called the normalized exponential function, has the following form:

$$f(x_1, \ldots, x_n) = \frac{1}{\sum_{i=1}^n e^{x_i}} [e^{x_1}, \ldots, e^{x_n}], \forall (x_1, \ldots, x_n) \in \mathbb{R}^n.$$

The softmax function has the properties that, for every $x \in \mathbb{R}^n$, each element of the output is non-negative, and the summation of all is equal to 1. Therefore it is often used to model a probability distribution. The softmax activation function is usually the right choice for the output layer of a classification problem.

5.1.3.6   *Identity function*

The identity function always returns the same value as the input, i.e.,

$$f(x) = x, \forall x \in \mathbb{R}.$$

The output of most activation functions discussed above is in a bounded range, which may not apply to regression problems where the output value is a continuous variable in $\mathbb{R}$. Therefore the identity activation function is usually used in the linear output layer of regression problems so that the network can learn a target value in any range.

5.1.4 Tensor

A tensor is an important concept in neural networks. All computation in neural networks boils down to tensor operations. Its rigorous mathematical definition is not straightforward to understand at first glance.[3] We provide an intuitive explanation for the tensor. You can imagine that a tensor is formed by a number of stackings over different axes. Figure 5.4 depicts the different dimensions of tensors.

5.1.4.1 *Scalar*

The scalar is a 0D tensor and also a number. There is no axis for the scalar.

5.1.4.2 *Vector*

The vector is a 1D tensor and has one axis. A scalar stacks over the horizontal axis and forms a vector. One-dimensional tensors are useful in artificial neural networks as the input is usually a 1D tensor.

5.1.4.3 *Matrix*

The matrix is a 2D tensor and has two axes. A vector stacks over the vertical axis and forms a matrix. A mutli-dimensional time series is a 2D tensor of size (time dimension, feature dimension). The 2D tensor is often used in recurrent neural networks as the input type.

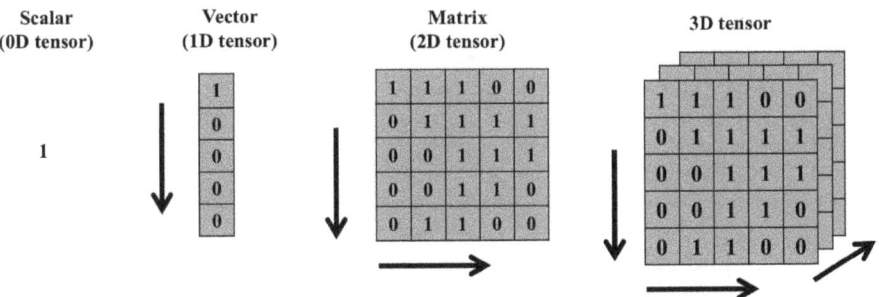

Figure 5.4. The illustration of different dimensions of tensors.

---

[3]You can find the rigorous definition of a tensor in [Maks *et al.* (1972)].

5.1.4.4 *N-dimensional tensor*

As we can image, if a matrix stacks over another new axis, we will have a 3D tensor. In general, an N-dimensional tensor can be thought of as an $N-1$ dimensional tensor stacking one by one over the new axis. The 3D tensor is especially important for convolutional neural networks as any color image is a 3D tensor.

5.2 Artificial Neural Network

The artificial neural network (ANN) is based on a collection of artificial neurons, which can transmit a signal from one artificial neuron to another. Let us start with the shallow neural network and explain its motivation from a mathematical perspective.

5.2.1 Shallow neural network

The shallow neural network is also called a 2-Layer ANN; its architecture is depicted in Figure 5.5.

The shallow neural network is a non-linear model to describe a mapping from the input space $\mathbb{R}^d$ to the output space $\mathbb{R}^e$. It is composed of one input layer $h^{(0)}$, one hidden layer $h^{(1)}$ and the output layer $h^{(2)}$. The lth layer is defined as a transformation from $\mathbb{R}^d$ to $\mathbb{R}^{n_l}$, where n_l denotes the number of neurons in the lth layer and $l \in \{1, 2\}$.

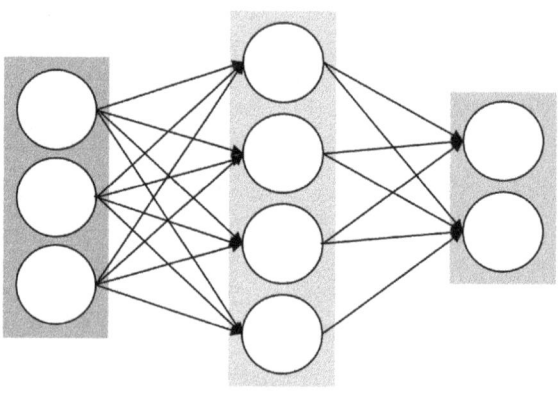

input layer   hidden layer   output layer

Figure 5.5. A 2-layer artificial neural network.

The shallow neural network is composed of three layers defined in a recursive way[4]:

(1) Input Layer ($h^{(0)} : \mathbb{R}^d \to \mathbb{R}^d$): $\forall x \in \mathbb{R}^d$,

$$x = (x^{(1)}, x^{(2)}, \ldots, x^{(d)}) \mapsto x,$$

where the input layer $h^{(0)}$ is the identity map, and neurons in the input layer are input x.

(2) Hidden Layer ($h^{(1)} : \mathbb{R}^d \to \mathbb{R}^{n_1}$): $\forall x \in \mathbb{R}^d$,

$$z^{(1)}(x) = W^{(1)}x + b^{(1)},$$
$$h^{(1)}(x) = \sigma_1(z^{(1)}),$$

where $W^{(1)}$ is an $n_1 \times d$ matrix of weights, $b^{(1)}$ is an n_1-dimensional vector, and σ_1 is called the activation function in the hidden layer and it is applied in an elementwise sense.

(3) Output Layer ($h^{(2)} : \mathbb{R}^d \to \mathbb{R}^e$): $\forall x \in \mathbb{R}^d$,

$$z^{(2)}(x) = W^{(2)}h^{(1)}(x) + b^{(2)},$$
$$h^{(2)}(x) = \sigma_2(z^{(2)}(x)),$$

where $W^{(2)}$ is an $n_2 \times n_1$ matrix of weights, $b^{(2)}$ is an n_2-dimensional vector, and σ_2 is called the activation function of the output layer, which is often chosen as the identity map for the regression problem, while the softmax function is typically used for classification problems.

Now let us explain why the shallow neural network with the sigmoid function $\sigma : x \mapsto \frac{1}{1+e^{-x}}$ is one of the natural choices for non-linear regression. We start with the Heaviside step function and explore the delicate connection between it and the sigmoid function.

Given a continuous function $f \in \mathcal{C}(J, \mathbb{R})$, the step function is a natural approximation, where J is a compact set of $\mathbb{R}$. On the other hand, any step function can be rewritten as a linear combination of Heaviside step functions. Therefore we have the following universality approximation theorem (UAT) for the linear function on Heaviside step functions (Theorem 5.1).

---

[4]The input layer is always the identity map for all the possible neural network. Although the shallow neural network has three layers, we call it a 2-layer Artificial neural network.

Theorem 5.1 (UAT for Step Function). *The finite sum of the form*

$$\sum_{i=1}^{n-1} C_n \mathbf{1}(t_i \leq x < t_{i+1})$$

is dense in $C(I_n)$. In other words, given any $f \in C(I_n)$ and $\varepsilon > 0$, there is a sum of the above form, denoted by F_C, such that for every $\varepsilon > 0$, there exists $(t_i)_{i=1}^n$ such that $t_1 < t_2 < \ldots < t_n$ and $t_i \in J, \forall i \in \{1, 2, \ldots, n\}$, and

$$\max_{x \in J} |f(x) - F_C(x)| \leq \varepsilon,$$

where $C := (C_i, t_i)_i$.

Now we extend the UAT from step functions to sigmoid functions. Let us consider a sequence of the parameterized sigmoid function, i.e.,

$$\sigma(x|\beta, a) := \frac{1}{1 + \exp(-\beta(x - a))}.$$

As we can see in Figure 5.6, the limit of $\sigma(x|\beta, a)$ converges to the Heaviside step function in the pointwise sense when β tends to infinity, i.e., for every $x \in \mathbb{R}/\{a\}$,

$$\lim_{\beta \uparrow \infty} \sigma(x|\beta, a) = \mathbf{1}(x > a). \tag{5.1}$$

Therefore it is further proved that the Heaviside step function can be uniformly approximated by a sequence of sigmoid functions by tuning the parameters β and a in Equation (5.1). Then if one adopts the approximation of the step function to model a continuous function, it is straightforward to think of a linear combination of sigmoid functions (not Heaviside step functions) as an alternative basis of continuous functions on the compact set of the input space. The benefit of choosing the sigmoid function rather than the Heaviside step function is its differentiability, which is essential for gradient descent based methods.

Theorem 5.2 (UAT for Shallow Neural Network [Gybenko (1989)]). *Let σ be any continuous discriminant function (e.g.,* *sigmoid function). The finite sum of the form*

$$\sum_{i=1}^{N} \alpha_j \sigma \left(\sum_{j=1} \beta_j^T x + \theta_j \right)$$

is dense in $C(I_n)$. In other words, given any $f \in C(I_n)$ and $\varepsilon > 0$, there is a sum of, $F_\Theta(x)$ of the above form, for which

$$\max_{x \in I_n} |F_\Theta(x) - f(x)| < \varepsilon,$$

where $\Theta := \{\alpha_j, \theta_j, \beta_j\}_{j=1}^N$.

5.2.2 Multi-layer ANN model architecture

The shallow neural network provides a universal model to approximate any continuous function $f : \mathbb{R}^d \to \mathbb{R}^e$. One may wonder why there is only one hidden layer and what would be gained by adding more hidden layers. This motivates us to propose the multi-layer artificial neural network.

Similarly to the shallow neural network, the multi-layer neural network is also composed of three types of layers, i.e., the input layer $h^{(0)}$, the hidden layer $(h^{(l)})_{l=1}^{L-1}$ and the output layer $h^{(L)}$. We still use the notations of the shallow neural network. Let $h^{(l)}$ denote the lth layer of the neural network for $l \in \{0, 1, \ldots, L\}$. Let n_l denote the number of neurons in the lth layer. $h^{(l)}$ is defined as a mapping from $\mathbb{R}^d$ to $\mathbb{R}^{n_l}$, where $l \in \{1, 2, \ldots, L\}$. By convention, $n_0 = d$.

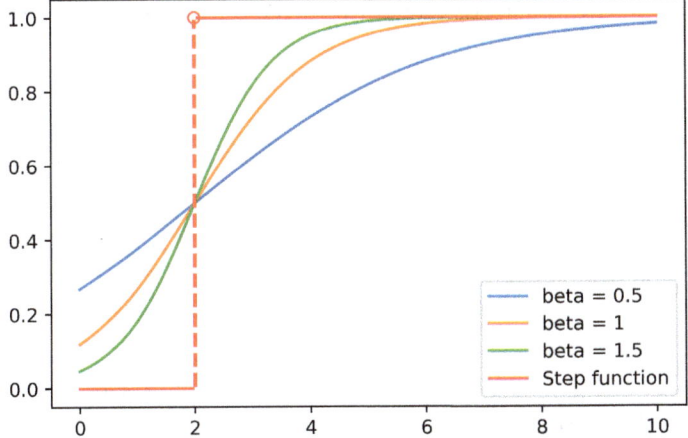

Figure 5.6. Sigmoid function with different parameters β.

- Input layer ($h^{(0)} : \mathbb{R}^d \to \mathbb{R}^{n_0}$):

$$h^{(0)}(x) = x.$$

- Hidden layer ($h^{(l)} : \mathbb{R}^d \to \mathbb{R}^{n_l}$): $\forall l \in \{1, 2, \ldots, L-1\}$.
- Output layer ($h^{(L)} : \mathbb{R}^d \to \mathbb{R}^e$).

$(h^{(l)})_{l=1}^{L}$ is defined in the following recursive way that for any $x \in \mathbb{R}^d$:

$$z^{(l+1)}(x) = W^{(l)}h^{(l)}(x) + b^{(l)}$$
$$h^{(l+1)}(x) = \sigma(z^{(l+1)}(x)),$$

where $W^{(l)}$ is an $n_{l+1} \times n_l$ matrix, $b^{(l)}$ is an n_{l+1} dimensional vector, σ is an activation function, e.g., the sigmoid function $\sigma(x) = \frac{1}{1+e^{-x}}$. The parameter set is $\theta = (W^{(l)}, b^{(l)})_{l=1}^{L}$.

Let us take a close look at the jth neuron in the $(l+1)^{th}$ layer, which is depicted in Table 5.1. For any given $x \in \mathbb{R}^d$, $z_j^{(l+1)}$ and $h_j^{(l+1)}$ are given by

$$z_j^{(l+1)} = \sum_{i=1}^{n_l} W_{i,j}^{(l)}h_i^{(l)} + b_j^{(l)}$$
$$h_j^{(l+1)} = \sigma(z_j^{(l+1)}),$$

where $W_{i,j}^{(l)}$ is the weight from incoming node i in layer l to output node j in layer $l+1$, and $b_j^{(l)}$ is the bias term to the output node j in layer $l+1$.

Table 5.1. Building block of the ANN.

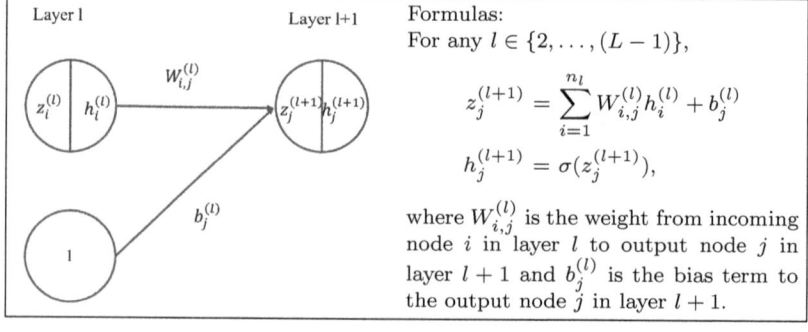

Table 5.2. Model of the multi-Layer ANN.

- Input layer ($h^{(0)} : \mathbb{R}^d \to \mathbb{R}^{n_0}$): $h^{(0)}(x) = x$.
- Hidden layer ($h^{(l)} : \mathbb{R}^d \to \mathbb{R}^{n_l}$): $\forall l \in \{1, 2, \ldots, L-1\}$.
- Output layer ($h^{(L)} : \mathbb{R}^d \to \mathbb{R}^e$):

  $$z^{(l+1)} = W^{(l)}h^{(l)} + b^{(l)},$$
  $$h^{(l+1)} = \sigma(z^{(l+1)}),$$

  where the activation function σ of hidden layers is usually non-linear, e.g., the sigmoid function $\sigma(x) = \frac{1}{1+e^{-x}}$. The parameter set is $\theta = (W^{(l)}, b^{(l)})_{l=1}^{L}$.

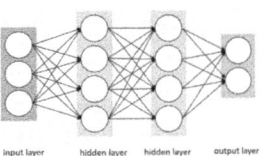

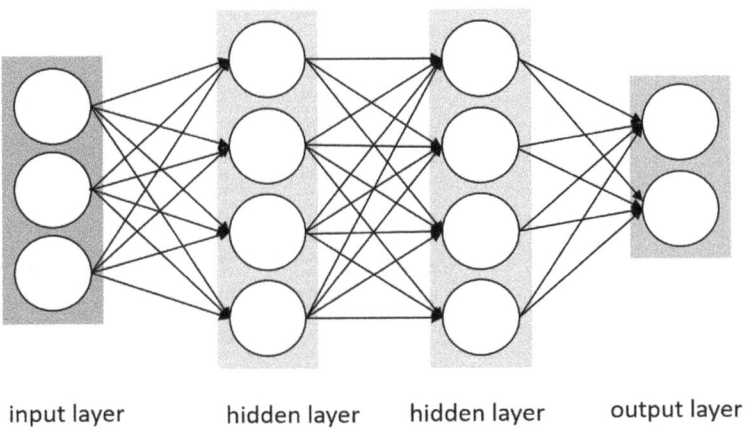

input layer · · · · · hidden layer · · · · · hidden layer · · · · · output layer

Figure 5.7. A regular three-layer ANN.

Sometimes the ANN is also called a fully connected neural network because all weights are connected. The model of the multi-layer ANN is summarized in Table 5.2. An example of the three-layer ANN architecture is given in Figure 5.7.

5.2.3 Optimization

For ANN models, there is no closed formula for the optimal parameters, and thus numerical optimization is required. As we discussed in Chapter 2, the standard optimization methods are mostly based on gradient descent, which requires the efficient evaluation of gradients. In our case, the training of ANN models also relies on the gradient descent based method, which searches for optimal parameters to minimize the loss function. Here the gradient calculation can be efficiently evaluated using the so-called *backpropagation* algorithm. In this subsection, we explain backpropagation algorithm step by step.

We still start with the loss function $L_\theta(\mathcal{D})$, which is usually in the additive form, i.e.,

$$L_\theta(\mathcal{D}) = \sum_{i=1}^{N} Q_\theta(x_i, y_i) = \sum_{i=1}^{N} Q(h^{(L)}(x_i) - y_i),$$

where θ is the set of model parameters of the multi-layer ANN. The gradient of $L_\theta(\mathcal{D})$ is given by the following equation by the chain rule:

$$\nabla_\theta L_\theta(\mathcal{D}) = \sum_{i=1}^{N} \nabla_\theta Q(h^{(L)}(x_i) - y_i)$$

$$= \sum_{i=1}^{N} Q'(h^{(L)}(x_i) - y_i) \nabla_\theta h^{(L)}(x_i),$$

provided that Q is a differentiable function.

The original problem has been reduced to how to compute the derivative of the output layer $h^{(L)}(x)$. You may think that this is not a difficult question as we have the symbolic representation of the output layer, and we can compute the derivative with respect to each parameter one by one. However, in practice, a neural network is usually built with millions of parameters; hence computing the derivatives one by one is very inefficient.

Therefore we do need to take advantage of the recursive structure of the neural network in order to design an efficient algorithm for the derivatives $\nabla_\theta h^{(L)}(x)$. One commonly used algorithm is the backpropagation algorithm, which exploits the hierarchical structure of the ANN and the chain rule. The procedure of the backpropagation algorithm to compute the gradient of $L_\theta(\mathcal{D})$ is composed of two phases:

(1) forward phase;
(2) backward phase.

In the forward phase, we evaluate all the neurons for the given input $x \in \mathbb{R}^d$ and fixed model weights θ. For each $l \in \{1, \ldots, L-1\}$, compute

$$z^{(l+1)}(x) = W^{(l)}h^{(l)}(x) + b^{(l)},$$
$$h^{(l+1)}(x) = \sigma(z^{(l+1)}(x)).$$

Figure 5.8 visualizes the flow chart of the forward phase.

In the backward phase, we backward compute the derivative of $Q_\theta(x, y)$ with respect to the weights θ in the lth layer for $l \in \{L-1, \ldots, 1\}$. We derive the recursion of derivatives between two consecutive layers using the chain rule.[5]

Let us recall that the model parameters of the ANN are $\theta = (\theta^{(l)})_{l \in \{1, \ldots, L\}}$, where $\theta^{(l)} = (W^{(l)}, b^{(l)})$. The key observation is that $Q_\theta(x, y)$ is a function depending on $z^{(l+1)}(x)$, $W^{(l+1)}, b^{(l+1)}, \ldots, W^{(L-1)}, b^{(L-1)}$ and y, where only $z^{(l+1)}(x)$ depends on $W^{(l)}$ and $b^{(l)}$. Thus it makes sense to

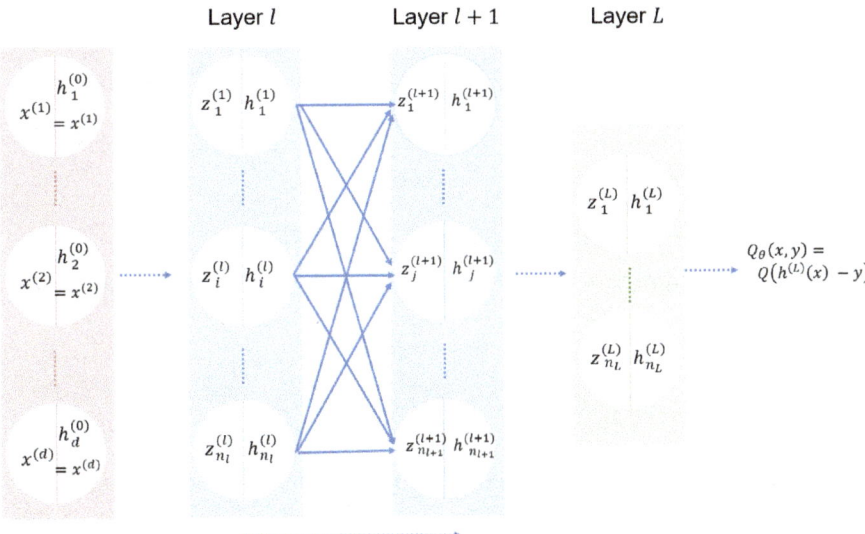

Figure 5.8. Backpropagation forward phase.

---

[5]The link `https://brilliant.org/wiki/backpropagation/` provides a good reference for the backpropagation algorithm.

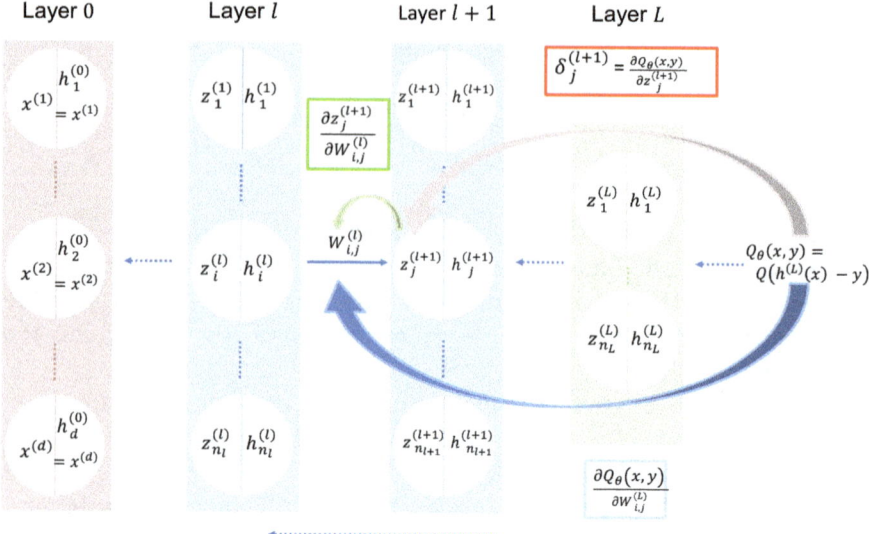

Figure 5.9. Illustration of recursive computation for $\partial_{W_{i,j}^{(l)}} Q_\theta(x, y)$.

regard the partial derivative of $Q_\theta(x, y)$ with respect to $z^{(l+1)}$ as the important intermediate variable,

$$\delta_i^{(l)} := \partial_{z_i^{(l)}} Q_\theta(x, y).$$

Furthermore, this yields a simpler formulae for $\partial_{W_{i,j}^{(l)}} Q_\theta(x, y)$ and $\partial_{b^{(l)}} Q_\theta(x, y)$ by applying the chain rule shown in Figure 5.9. In formulae,

$$\partial_{W_{i,j}^{(l)}} Q_\theta(x) = \partial_{W_{i,j}^{(l)}} z_j^{(l+1)} \cdot \partial_{z_j^{(l+1)}} Q_\theta(x) = \left(\partial_{W_{i,j}^{(l)}} z_j^{(l+1)} \right) \delta_j^{(l+1)},$$

$$\partial_{b_i^{(l)}} Q_\theta(x) = \partial_{b_i^{(l)}} z_j^{(l+1)} \cdot \partial_{z_j^{(l+1)}} Q_\theta(x) = \left(\partial_{b_i^{(l)}} z_j^{(l+1)} \right) \delta_j^{(l+1)}.$$

Now the original problem is reduced to computing $\partial_{W_{i,j}^{(l)}} z_j^{(l+1)}$, $\partial_{b_i^{(l)}} z_j^{(l+1)}$ and $\delta_i^{(l)}$,

$$\delta_i^{(l)} = \sum_{j=1}^{n_{l+1}} \delta_j^{(l+1)} \partial_{z_i^{(l)}} z_j^{(l+1)}.$$

Computation of $\partial_{W_{i,j}^{(l)}} z_j^{(l+1)}$ is straightforward, and the result is given in the following lemma:

Lemma 5.1 (Computation of $\partial_{W_{i,j}^{(l)}} z_j^{(l+1)}$).

$$\partial_{W_{i,j}^{(l)}} z_j^{(l+1)} = \partial_{W_{ij}^{(l)}} \left(\sum_k W_{k,j}^{(l)} h_k^{(l)} + b_j^{(l)} \right) = h_i^{(l)}$$

$$\partial_{b_i^{(l)}} z_j^{(l+1)} = \partial_{b_i^{(l)}} \left(\sum_k W_{k,j}^{(l)} h_k^{(l)} + b_j^{(l)} \right) = 1$$

$$\partial_{z_i^{(l)}} z_j^{(l+1)} = \partial_{z_i^{(l)}} \left(\sum_k W_{k,j}^{(l)} h_k^{(l)} + b_j^{(l)} \right) = \partial_{z_i^{(l)}} \left(W_{i,j}^{(l)} h_i^{(l)} \right)$$

$$= W_{i,j}^{(l)} \partial_{z_i^{(l)}} \left(\sigma(z_i^{(l)}) \right) = W_{i,j}^{(l)} \sigma' \left(z_i^{(l)} \right).$$

The recursive structure of δ_i^l can be well exploited by the chain rule depicted in Figure 5.10.

Lemma 5.2 (Computation of $\delta_i^{(l)} = \partial_{z_i^{(l)}} Q_\theta(x, y)$).

$$\delta_i^{(l)} = \sum_j \partial_{z_i^{(l)}} z_j^{(l+1)} \delta_j^{(l+1)} = \sum_j \sigma' \left(z_i^{(l)} \right) W_{i,j}^{(l)} \delta_j^{(l+1)} = \sigma' \left(z_i^{(l)} \right) \left(W^{(l)} \delta^{(l+1)} \right)_i.$$

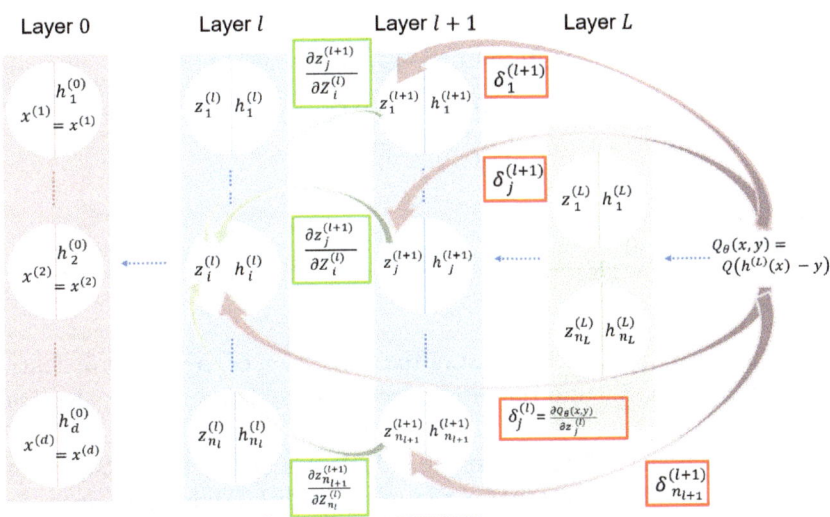

Figure 5.10. Illustration of recursive computation for $\delta_i^{(l)}$.

Table 5.3. Core idea behind the backpropagation algorithm.

Goal: To compute $\nabla Q_\theta(x, y)$, i.e.,

$$\partial_{W_{i,j}^{(l)}} Q_\theta(x, y) \text{ and } \partial_{b_i^{(l)}} Q_\theta(x,y),$$

where $Q_\theta(x, y) = Q(h^{(L)}(x) - y)$.
Idea: Recursion and chain rule.

$$Q_\theta(x, y) = q(z^{(l+1)}(x), W^{(l+1)}, b^{(l+1)}, \ldots, W^{(L-1)}, b^{(L-1)}, y),$$

where only $z^{(l+1)}(x)$ depends on parameter $W^{(l)}$ and $b^{(l)}$.

Algorithm 8: Backpropagation Algorithm

1: **Input:** θ and (x, y).
2: **Forward Phase:**
3: **for** $l = 1 : L - 1$ **do**
4:    Compute $z^{(l+1)}(x)$ and $h^{(l+1)}$

$$z^{(l+1)}(x) = W^{(l)} h^{(l)}(x) + b^{(l)},$$
$$h^{(l+1)}(x) = \sigma(z^{(l+1)}(x)).$$

5: **end for**
6: **Backward Phase:**
7: For $l = L$, $\delta^{(L)} = \partial_{z^{(L)}} Q_\theta(x, y) = \partial_{z^{(L)}} Q(h^{(L)}(x) - y) = Q'(h^{(L)}(x) - y)\sigma'(z^{(L)}(x));$
8: **for** $l = (L - 1) : 1$ **do**
9:    Compute $\delta^{(l)}$, $\partial_{W_{i,j}^{(l)}} Q_\theta(x, y)$ and $\partial_{b_j^{(l)}} Q_\theta(x, y)$

$$\delta^{(l)} = \sigma'(z^{(l+1)}(x)) \odot (W^{(l)} \delta^{(l+1)}),$$
$$\partial_{W_{i,j}^{(l)}} Q_\theta(x, y) = h_i^{(l)} \delta_j^{(l+1)},$$
$$\partial_{b_j^{(l)}} Q_\theta(x, y) = \delta_j^{(l+1)}.$$

10: **end for**
11: **return** $\nabla_\theta Q_\theta(x, y)$

Let us summarize the key idea of the derivation of the backpropagation algorithm in Table 5.3 and the procedure of the backpropagation algorithm in Algorithm 8.

After discussion of the gradient calculation using the backpropagation algorithm, we can use it for the mini-batch GD method (covered in Section 2.1.3.5), which is a popular numerical optimization method and commonly used for parameter optimization in neural network models. The

---

Algorithm 9: Mini-Batch GD

1: **Input:** $\mathcal{D} = \{(x_i, y_i)\}_{i=1}^N$, $\theta_0, N_{\text{epoches}}$ and m (the size of mini-batch).
2: Initialize $\theta = \theta_0$.
3: **for** $n = 1 : N_{\text{epoches}}$ **do**
4:     Randomly reshuffle the whole dataset $\mathcal{D} = \{(x_i, y_i)\}_{i=1}^N$;
5:     **for** $q = 1 : N_{\text{batches}}$ **do**

6:
$$\theta = \theta - \frac{1}{m}\sum_{i=1}^{m}\eta\nabla_\theta Q_{\theta_n}(x_{(q-1)m+i}, y_{(q-1)m+i}),$$

    where $N_{\text{batches}} = \lceil N/m \rceil$.
7:     **end for**
8: **end for**
9: **return** θ

---

mini-batch GD algorithm and its variants may be the most popular one, as it is a hybrid method of both GBD and SGD that can be used in empirical data sets of decent size or even large scale. Mini-batch GD is summarized in Algorithm 9.

5.2.4 Numerical example: MNIST digit recognition

In this subsection, let us consider the classification problem on handwritten digits. Our data comes from the MNIST dataset.[6] The MNIST dataset is the most popular dataset in digit recognition, and is composed of handwritten images of the numbers 0–9. It contains 60,000 training images and 10,000 testing images taken from American Census Bureau employees and American high school students. Samples of MNIST dataset are provided in Figure 5.11.

Before applying the ANN on the MNIST dataset, we need to pre-process the data to prepare it for the ANN (see Listing 5.7). After that, we implement the ANN for digit classification using Keras and the code for this implementation is given in Listing 5.8. You can also find other informative python demo code samples on the following website.[7]

We provide the codes for visualizing the fitting results in Listing 5.9. The fitting result is given in Figures 5.12 and 5.13. The confusion matrix of the calibrated shallow neural network model is given in Figure 5.14. It shows that the most confusing digit pair is $(3, 8)$, based on the confusion

---

[6]http://yann.lecun.com/exdb/mnist/.
[7]http://parneetk.github.io/blog/neural-networks-in-keras/.

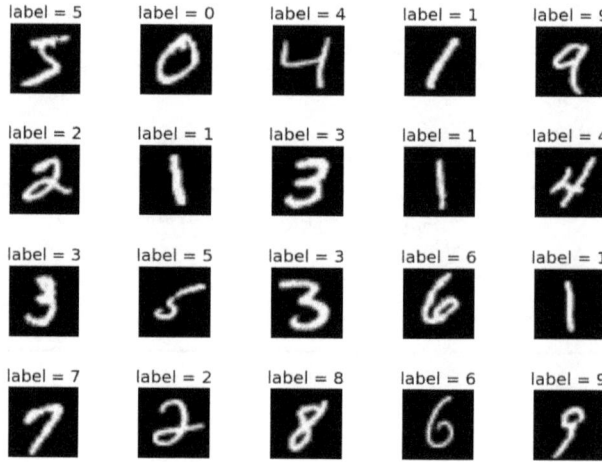

Figure 5.11.   Samples of MNIST dataset.

Source: http://corochann.com/wp-content/uploads/2017/02/mnist_plot.png.

```
1    import numpy as np
2    from keras.datasets import mnist
3    from keras.utils import to_categorical
4
5    # Import MNIST dataset
6    (x_mnist_train, y_mnist_train), (x_mnist_test, y_mnist_test) =
     ↪   mnist.load_data()
7
8    # Reshape input image data from matrices to vectors
9    [n_samples_train, width, height] = np.shape(x_mnist_train)
10   [n_samples_test, width, height] = np.shape(x_mnist_test)
11   x_train = x_mnist_train.reshape([n_samples_train,
     ↪   width*height]).astype('float32')
12   x_test = x_mnist_test.reshape([n_samples_test,
     ↪   width*height]).astype('float32')
13
14   # Normalize the dataset
15   x_train /= 255
16   x_test /= 255
17
18   # Convert class vectors to one-hot vector matrices
19   num_classes = 10
20   y_train = to_categorical(y_mnist_train, num_classes)
21   y_test = to_categorical(y_mnist_test, num_classes)
```

Listing 5.7. Pre-process MNIST dataset.

```
1   from tensorflow.keras.models import Sequential
2   from tensorflow.keras.layers import Dense
3   from tensorflow.keras.optimizers import SGD
4
5   def shallow_NN_model(n_hidden_neurons):
6       model = Sequential()
7       # Add a hidden layer with ReLU activation function
8       model.add(Dense(n_hidden_neurons, activation='relu', input_dim=784))
9       # Add an output layer with softmax activation function for
    ↪   multi-classification problem
10      model.add(Dense(10, activation = 'softmax'))
11      # Here we use the general setting of SGD in Keras
12      sgd = SGD(lr=0.1, decay=1e-6, momentum=0.9, nesterov=True)
13      model.compile(loss='categorical_crossentropy', optimizer=sgd,
    ↪   metrics=['accuracy'])
14      return model
15
16  epochs = 100
17  batch  = 256
18  n_hidden_neurons = 50
19  model = shallow_NN_model(n_hidden_neurons)
20  hist_ANN = model.fit(x_train, y_train, epochs=epochs, batch_size=batch,
    ↪   validation_data=(x_test, y_test), verbose=2)
```

Listing 5.8. Implement ANN for digit classification.

```
1   import matplotlib.pyplot as plt
2
3   # Define a function to plot the loss and accuracy
4   def plot_hist_var(hist, str_var):
5   plt.figure()
6   plt.plot(hist.history[str_var], 'b', linewidth=1.5)
7   plt.plot(hist.history['val_'+str_var], 'r', linewidth=1.5)
8   plt.legend(['Training set','Testing set'])
9   plt.xlabel('Epochs')
10  plt.ylabel(str_var[0].upper()+str_var[1:])
11
12  plot_hist_var(hist_ANN, 'loss')
13  plot_hist_var(hist_ANN, 'accuracy')
```

Listing 5.9. Plot the fitting result.

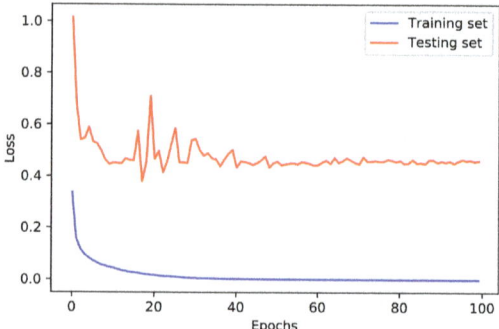

Figure 5.12. The loss of using ANN for MNIST digit classification on the training and testing set.

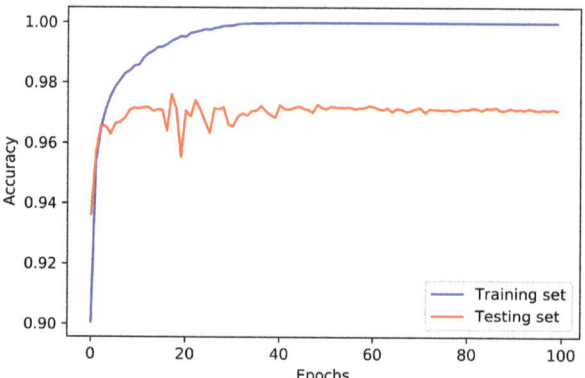

Figure 5.13. The accuracy of using ANN for MNIST digit classification on the training and testing set.

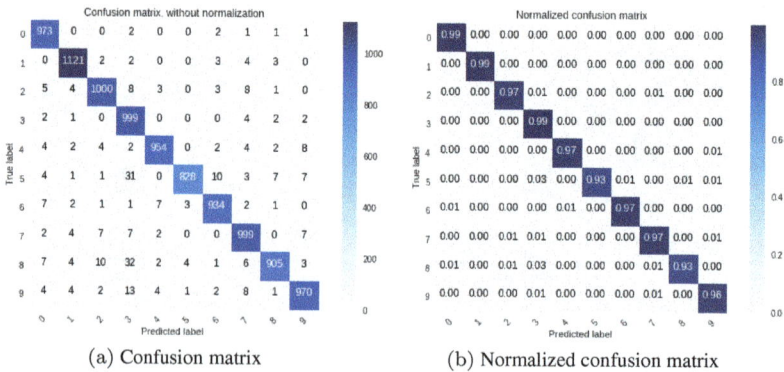

(a) Confusion matrix   (b) Normalized confusion matrix

Figure 5.14. The (normalized) confusion matrix of the estimated shallow NN for MNIST digit classification.

matrix. The accuracy that using a shallow neural network on the MINST dataset can achieve is around 97%.

5.3 Convolutional Neural Network

The Convolutional Neural Network (CNN) is a category of neural network that has proven very effective in image related machine learning tasks. Compared with ANNs, CNNs can retain the original spatial information of image data to the greatest extent. CNN has been successfully applied to solve various real-world problems, for example, face recognition, scene labeling, image classification, action recognition, human pose estimation and document analysis. Besides, it has been proved that a CNN can achieve the best accuracy in the field of speech recognition and text classification for natural language processing. For a comprehensive summary of the applications of CNN, the reader is referred to [Bhandare *et al.* (2016)].

5.3.1 Introduction and motivation

CNN is inspired by the structure of the visual system, and was proposed by [Hubel and Wiesel (1962)]. This study suggests that cells in the cat's primary visual cortex are sensitive to small sub-regions of the visual field called a receptive field. CNN adopts a similar network structure, where there are two new types of layers, called convolutional layers and pooling layers. In the convolutional layer, the filter (usually smaller than the input) *convolves* with every receptive field over the original input volume and produces a feature map. In the pooling layer, the essential information is extracted while the other drops out through different pooling methods. This kind of layer structure makes CNN effectively learn the essential information of different sub-regions in the input.

Two unique characteristics of CNN are *local connectivity* and *shared parameters*, which reduce the vast number of model parameters. CNN is very similar to ANN except for the connectivity and the spatial structure of the input data. For example, if we have a grayscale image dataset with size $(28, 28, 1)$, we can directly apply CNN to the original data. As for ANN, we have to "flatten" the original image to be $(784, 1)$. Therefore ANN overlooks the shape of the input image and fails to achieve good accuracy for complicated image recognition tasks. Besides this, the layers in CNN are not fully connected: only a subset of the neurons in one layer are connected with neurons in other layers.

CNN and ANN also have many common characteristics. In CNN, the whole network exhibits a hierarchical structure of layers to model an input-output map: from the raw image pixels on one end to class scores at the other. Neurons in the same layer don't have direct connections; however, there is a direct dependence between two consecutive layers. Moreover, to use a CNN to train a model, one still needs to specify a loss function and use stochastic gradient descent to numerically estimate the optimal model parameters—all the tricks we developed for learning ANN still apply.

The LeNet Architecture (the 1990s) is one of the very first convolutional neural networks and propelled the field of deep learning [LeCun *et al.* (2015b)]. This pioneering work by Yann LeCun is named LeNet-5 after many previous successful iterations since 1988. At that time, the LeNet architecture was mainly used for character recognition tasks such as reading zip codes, digits, etc. Despite all these advantages, due to the lack of large training datasets and computational power at that time, LeNet-5 did not perform well on complex problems such as video classification. Thanks to the advent of GPUs and their use in machine learning [Steinkraus *et al.* (2005)], the field of CNNs has gone through a renaissance. In the recent development of deep learning for image recognition, new architectures have been proposed such as AlexNet [Krizhevsky *et al.* (2012)], VGGNet [Simonyan and Zisserman (2014)], GoogleNet [Szegedy *et al.* (2015)] and ResNet [He *et al.* (2016)]. These new networks are improvements over LeNet and demonstrate significant improvement in image classification tasks. Still, most of them use the main concepts introduced by LeNet and are easier to understand if you have a clear understanding of the former. In the following, we explain the main components of the convolutional neural network, which will help you to understand LeNet and its variants.

5.3.2   Problem setting and image data

What is image data? Essentially an image is a matrix of pixel values. As we can see in Figure 5.15, a grayscale image can be represented as a two-dimensional tensor (matrix) whose elements range in value from 0 to 255. A grayscale image (black and white image) with width W and height H can be represented as a matrix of size (W, H).

A color image has one more dimension, i.e., the depth dimension, which denotes the number of channels. A color image has three channels, i.e., red, green and blue (RGB for short). Therefore a color image with width W and height H can be well represented by a three-dimensional tensor of size $(W, H, 3)$. Figure 5.16 shows an example of a color image.

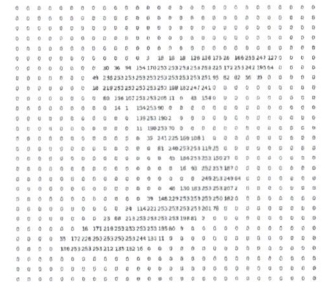

(a) A matrix of pixel values   (b) A sample of handwritten digit 5

Figure 5.15. Sample from the MNIST dataset.

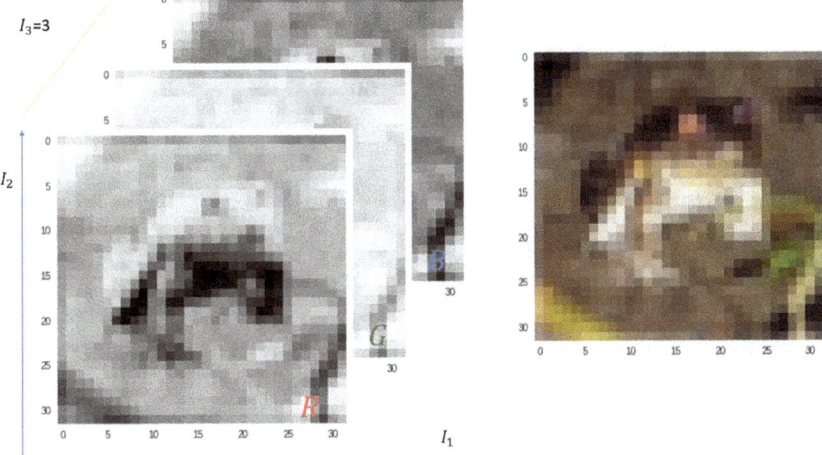

Figure 5.16. An image can be represented as 3-dimensional tensors (width, height, depth). In this example, the three pictures on the left denote three channels (RGB), which stack over the depth axis to form the color picture on the right.

The definition of image data in give in Definition 5.1. From now on, $\forall$ integers n, $[n]$ denotes the set $\{1, 2, \ldots, n\}$.

Definition 5.1 (Image Type Data). An image dataset of size $(W \times H)$ with d color channels is a mapping from $[W] \times [H] \to \mathbb{R}^d$ or it can be represented as a three-dimensional tensor $X = (X_{i,j,k})_{i \in [W], j \in [H], k \in [d]} \in \mathbb{R}^{W \times H \times d}$. We denote the space of all the mappings from $[W] \times [H] \to \mathbb{R}^d$ by $\mathcal{F}([W] \times [H], \mathbb{R}^d)$.

Remark 5.1. An image type dataset of size $(W \times H)$ with d dimensional color channels has three dimension (W, H, d), where W and H are the width and height of the image (spatial dimension), and d is the color channel dimension. The spatial dimension is different from the channel dimension as it has the physical meaning of closeness in space; this is not the case for the channel (feature) dimension.

5.3.3 Model

Like LeNet in Figure 5.17 a general convolutional neural network is mainly composed of three main types of layers:

- Convolutional layer.
- Pooling layer.
- Fully connected (Dense) layer.

In the following subsections, we discuss the convolutional layer and pooling layer in detail. Then we will return to the LeNet architecture and its application to the Cifar10 image recognition task.

5.3.3.1 *Convolutional layer*

The convolutional layer is a core building block of a CNN. The main purpose of a convolutional layer is to extract features from the input image efficiently, preserving the spatial relationship between pixels. The core idea is to learn image features using small squares of input data and share weights across all small patches throughout the image. The parameters of the convolutional layer consist of a set of learnable filters.[8] Every filter is defined over a spatially (along width and height) small region, but it extends through the full depth of the input volume.

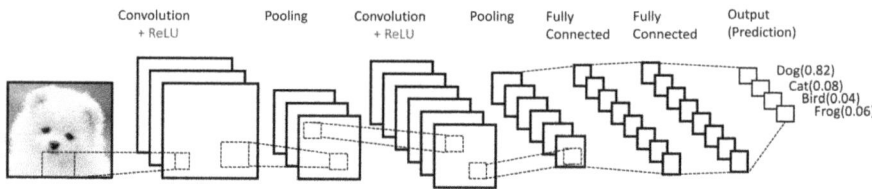

Figure 5.17.   LeNet architecture.

Source: Retrieved from https://images.app.goo.gl/3R73mCJSyUu3mCWx9.

---

[8]The filters are also called the convolutional kernel. We use the terminology 'filter' in this book.

1	1	1	0	0
0	1	1	1	0
0	0	1	1	1
0	0	1	1	0
0	1	1	0	0

1	0	1
0	1	0
1	0	1

(a) A grey-valued image input $(X_{i,j})_{i,j=1}^5$ (b) A 2D filter $(W_{i,j})_{i,j=1}^3$

Figure 5.18. An example of a gray-valued image and a filter of the convolutional layer.

For example, a typical filter on the first convolutional layer might have size $5 \times 5 \times 3$ (i.e., 5 pixels width and height, and the depth is 3 because images have 3 color channels). During the forward pass, we slide (more precisely, convolve) each filter across the width and height of the input volume. During the sliding process, the corresponding small region of the input is the receptive field. Then we compute dot products between the filter and the receptive field at each position. The sliding process produces a two-dimensional feature map, which gives the responses of that filter at every spatial location. We explain the detail of how the convolutional layer works and its mathematical definition in the following section. Interested readers may refer to the following websites for more information about CNN.[9,10]

Let us start with an example of a black and white image in Figure 5.18. As we discussed above, every gray-valued image can be considered as a matrix of pixel values. In general, for a grayscale image, pixel values range from 0 to 255. The green matrix in Figure 5.18a denoted by $(X_{i,j})_{i,j=1}^5$ is a special case where pixel values are only 0 and 1, which represents a 5×5 grayscale image.

Now we explain how the convolution computation works. In Figure 5.18b, the yellow matrix is a filter, denoted by $(W_{i,j})_{i,j=1}^3$. We slide this filter over the yellow image in Figure 5.18a with a 1 pixel stride. For every position, we compute the multiplication between the corresponding entries of the two matrices. Then sum the multiplication outputs to get a scalar, which forms a single element of the output matrix, called a feature

---

[9]https://ujjwalkarn.me/2016/08/11/intuitive-explanation-convnets/.
[10]http://cs231n.github.io/convolutional-networks/.

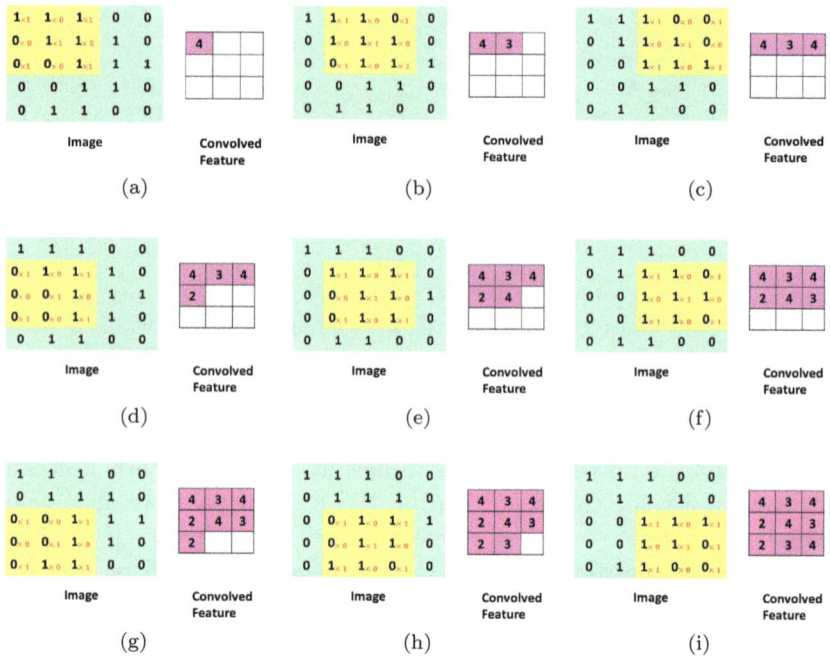

Figure 5.19. An example of computing one convolutional layer with a black and white image.

map. Figure 5.19 describes the whole procedure, and the feature map is in pink.

For example, in Figure 5.19a, the first element in pink is obtained by applying the summation of the Hadamard (elementwise) multiplication of $(X_{i,j})_{i,j=1}^3$ and $(W_{i,j})_{i,j=1}^3$. Note that the first element in pink is only dependent on a sub-region of the original image, which is the receptive field (the yellow part in Figure 5.19a).

$$(X_{i,j})_{i,j=1}^3 \cdot (W_{i,j})_{i,j=1}^3 = \begin{bmatrix} 1 & 1 & 1 \\ 0 & 1 & 1 \\ 0 & 0 & 1 \end{bmatrix} \cdot \begin{bmatrix} 1 & 0 & 1 \\ 0 & 1 & 0 \\ 1 & 0 & 1 \end{bmatrix} := \sum_{i,j=1}^3 X_{i,j} W_{i,j} = 4.$$

In the above example, the image is of size $(5, 5)$, the kernel size is $(3, 3)$ and the step of sliding the filter (called the stride) is $(1, 1)$. In general, suppose that we have an image of size (w_1, w_2), a kernel of size (f_1, f_2) and a stride of size (s_1, s_2). The last sliding window may exceed the region of the image. To get around this problem, zero padding is a common operation in the convolution layer, which refers to adding zeros to the input volume around the image boundary. Let (p_1, p_2) denote the dimension of the zero padding; it holds that $\forall i \in \{1, 2\}$,

$$w_i + p_i = f_i + \tilde{w}_i s_i,$$

where $\tilde{w}_i$ is the largest integer such that $w_i \geq f_i + (n-1)s_i$. Thus the output of the convolutional operator has the size $(\tilde{w}_1, \tilde{w}_2)$, which is given as follows:

$$\tilde{w}_i = \frac{w_i - f_i + p_i}{s_i} + 1, \forall i \in \{1, 2\}. \tag{5.2}$$

The computation of the convolutional operator on a color image is similar to that of the gray-valued image. We still multiply the filter and the receptive field to get the feature map. However, they are both 3D tensors in the color image instead of 2D tensors for the black and white image. In this case, the spatial size of the filter is usually smaller than that of the input image; but the depth dimension of the filter should be the same as that of the input image. The filters are shared parameters to obtain the feature map when sliding the receptive fields during the convolution.

Remark 5.2 (*Shared Parameters*). The color image can be viewed as a stack of matrices $(X^i)_{i=1}^3$ along the depth, where X^i represents the image projected onto the ith color channel. The filter of a color image is also a $3D$ tensor—a stack of matrices $(W^i)_{i=1}^3$, where W^i are matrices of the same size for $i \in \{1, 2, 3\}$. It is noted that the dimension of W^i is smaller than that of X^i. But X and W have the same depth. The convolution on the 3D tensor image is to sum up the convolution results of the 2D tensor image of each color channel X^i and the filter slice W^i, where the summation is taken over i from 1 to 3. Figure 5.20 illustrates that the filter with the orange boundary is of the 3D tensor type. The three black-and-white images represent the original image projected onto each color channel, and the corresponding filter slice is depicted under each image as a box with an orange boundary. Let us start with the receptive field in the same spatial region with the blue boundary. For each color channel, we compute the convolution result of each black-white image within the blue boundary with the filter slice. We sum them up along all the RGB depth and produce the corresponding blue element in the output neuron. Similarly, sliding the receptive field marked with the green boundary by one stride produces the corresponding green element in the output neuron. After sliding the receptive field throughout

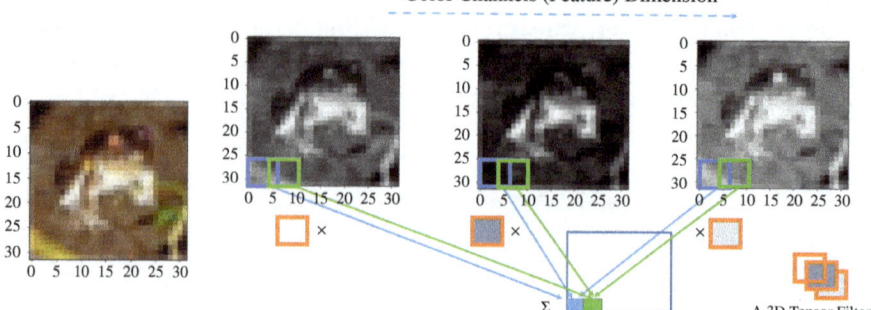

Figure 5.20. Visualization of convolutional layer computation with color image input.

the whole image, we obtain a matrix output as a feature map. During this process, the convolution operator shares the parameters through the different receptive fields.

For the color image, the output of one filter is still a 2D tensor, with the same dimension as that of the gray-valued image. Now suppose that we have multiple filters, given one input image dataset; the output for the different filters stacks over the depth axis and forms a 3D tensor.

After working on concrete examples of computing the convolved features on both a black-white image and a color image, let us abstract this and offer a formal definition of the convolutional operator and convolutional layer. For ease of discussion, we introduce some notations.

Let $\tilde{n}_1 + [\tilde{n}_2] := (\tilde{n}_1 + 1, \tilde{n}_1 + 2, \ldots, \tilde{n}_1 + \tilde{n}_2)$, $\forall \tilde{n}_1, \tilde{n}_2 \in \mathbb{N}^+$. For the convolutional operation, the spatial domain of the receptive field at step (x_1, x_2) with stride (s_1, s_2) and kernel size (f_1, f_2) is

$$((x_1 - 1)s_1 + [f_1]) \times (x_2 - 1)s_2 + [f_2]. \qquad (5.3)$$

Equation (5.3) can be rewritten as $s \circ (x - 1) + f$, where $\circ$ is the Hadamard product of two matrices.

Definition 5.2 (Convolutional Operator). Let $\mathcal{C}_W$ denote the convolutional operator of stride $s \in \mathbb{Z}^2$ mapping from $\mathcal{F}([w_1] \times [w_2], \mathbb{R}^d)$ to $\mathcal{F}([\tilde{w}_1] \times [\tilde{w}_2], \mathbb{R}^{\tilde{d}})$, where W is a filter of a 4D tensor of size $(f_1, f_2, d, \tilde{d})$ if the following condition holds. Assume that for $i = 1, 2$, $s_i \in [w_i]$. For any $h \in \mathcal{F}([w_1] \times [w_2], \mathbb{R}^d)$, $\forall x \in [\tilde{w}_1] \times [\tilde{w}_2]$,

$$\mathcal{C}_W(h)(x) = \int_{\mathbb{Z} \times \mathbb{Z}} h(s \circ (x - 1) + f - y) \cdot \Phi_W(y) dy$$

$$:= h(s \circ (x - 1) + f) * \Phi_W,$$

where ∘ is the Hadamard product of two matrices. The kernel/filter function Φ_W is given as follows: $\forall x = (x_1, x_2) \in \mathbb{Z}^d$,

$$\Phi_W(x) = \mathbf{1}(x_1 \in [f_1], x_2 \in [f_2]) W_{x_1, x_2} \in \mathbb{R}^{d \times \tilde{d}}.$$

$(\tilde{w}_1, \tilde{w}_2)$ is induced by Equation (5.2).[11]

Definition 5.3 (Convolutional Layer).
Let $h^{(l)} : \mathcal{F}([W_0] \times [H_0], \mathbb{R}^{d_0}) \to \mathcal{F}([W_l] \times [H_l], \mathbb{R}^{d_l})$ be the lth layer of the neural network. Let $W \in \mathcal{F}([f_1] \times [f_2], \mathbb{R}^{d_{l-1} \times d_l})$ be a filter of size (f_1, f_2, d_{l-1}, d_l):

$$h^{(l)} = \sigma(\mathcal{C}_W(h^{(l-1)})),$$

where σ is the activation function and $\mathcal{C}_W$ is the convolution operator of stride $s \in \mathbb{Z}^2$ mapping from $\mathcal{F}([W_{l-1}] \times [H_{l-1}], \mathbb{R}^{d_{l-1}})$ to $\mathcal{F}([W_l] \times [H_l], \mathbb{R}^{d_l})$. (W_l, H_l) is induced by Equation (5.2) with the spatial dimension of the input image being (W_{l-1}, H_{l-1}), the kernel size (f_1, f_2) and the stride s. Then we call $h^{(l)}$ a convolutional layer with stride s and filter size (f_1, f_2, d_{l-1}, d_l).

Remark 5.3. The dimension of trainable parameters of the convolutional neural network is that of the filter, i.e., $f_1 \times f_2 \times d \times \tilde{d}$, which is independent of the spatial size of the input image.

It is important to note that filters act as feature detectors from the original input image. It is evident from Figure 5.20 that the set of convolved features/feature map is a function with respect to the filters.

The layer's parameters consist of a set of filters, which have a small spatial size but extend through the full depth of the input volume. During the forward pass, each filter is convolved across the width and height of the input volume, computing the dot product between the entries of the filter and the input and producing a two-dimensional activation map of that filter.

The extent of the connectivity along the depth axis is always equal to the depth of the input volume. It is important to emphasize again this asymmetry in how we treat the spatial dimension (width and height) and the depth dimension: the connections are local in space (along width and height), but always full along the entire depth of the input volume (see Figure 5.21).

---

[11]The convolution operator of two functions f and g is defined as follows: for all $t \in \mathbb{R}$, $f * g(t) = \int_s f(t-s)g(s)ds$.

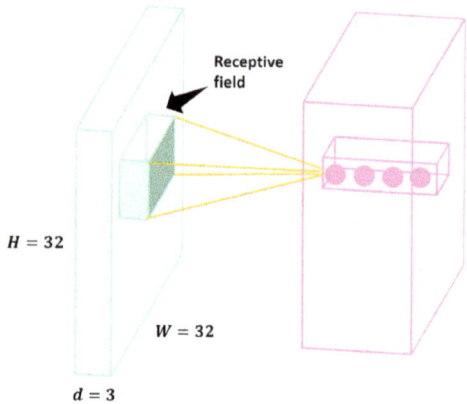

Figure 5.21. An example input volume in red (a 32 × 32 × 3 image sample) from the Cifar10 dataset and an example volume of neurons in the first convolutional layer. Each neuron in the convolutional layer is connected only to a local region in the input volume spatially, but to the full depth (i.e., all color channels). Note that there are multiple neurons (four in this example) along the depth, all looking at the same region in the input.

Remark 5.4 (*Local Connectivity*). When dealing with high-dimensional image inputs, it is impractical to use the ANN—even just a two-layer ANN. For example, a square image of moderate size, such as 100×100, needs $100^2 \times 3$ weights for a single neuron in the hidden layers. Taking more neurons in the hidden layers and multiple layers into account, the vast number of model parameters of the ANN for image recognition tasks results in heavy computation for the training, which may be beyond standard computational capacity. Relative to the massive number of parameters, the sample size of the image is much too sparse, which would quickly lead to overfitting if using an ANN. Instead, the CNN proposes to only connect each neuron to a local region of the input volume and to share model parameters across different receptive fields. The spatial extent of this connectivity is a hyperparameter called the receptive field size of the neuron (equivalent to the filter size). The dimension of trainable parameters of the convolutional neural network is the size of the filter, which is typically far less than that of a dense layer.

5.3.3.2 *Pooling layer*

The pooling layer is another important layer in the CNN. Spatial pooling (also called subsampling or downsampling) reduces the dimensionality of each feature map but aims to capture the most important information. Spatial pooling can be of various types: max, average, sum, etc.

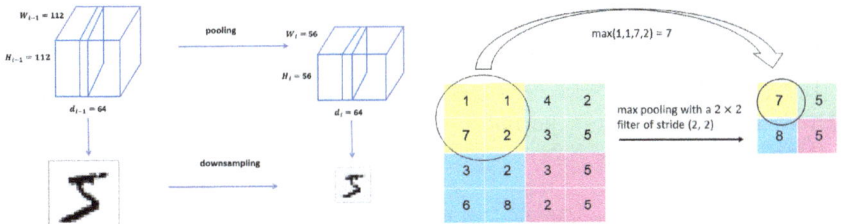

Figure 5.22. Example of the computation of the pooling layer.

Take max pooling as an example (see Figure 5.22). We define a spatial neighborhood (for example, a 2×2 window) of the feature map. Then we take the largest element within that window in turn to constitute a new matrix. The pooling layer performs a downsampling operation along the spatial dimensions, resulting in volume reduction. The spatial dimension of the image reduces from 224×224 to 112×112 in Figure 5.22.

In Figure 5.22, the pooling layer has stride $(2, 2)$ and kernel size $(2, 2)$. The receptive field at step $(1, 1)$, marked in yellow, takes values in the spatial domain $\{1, 2\} \times \{1, 2\}$. If moving one step to the right, which corresponds to step $(1, 2)$, the receptive filed marked in green has the spatial domain $\{1, 2\} \times \{3, 4\}$. In general, for the pooling operation, the spatial domain of the receptive field at step (x_1, x_2) with stride (s_1, s_2) and kernel size (f_1, f_2) is $((x - 1) \circ s + [f])$ (Equation (5.3)), similar to that of the convolutional layer.

To give a formal mathematical definition of the pooling operator and the pooling layer, we introduce the following notations. Let h be a mapping from $\mathbb{Z}^2$ to $\mathbb{R}^d$ and $S \subset \mathbb{Z}^2$. $h(S)$ is defined to be a set of $h(s)$ for all $s \in S$, i.e., $h(S) := \{h(s) | s \in S\}$. Let g denote a map $\mathbb{R}^{f_1 \times f_2} \to \mathbb{R}$. We define the induced map $\tilde{g} : h \in \mathbb{R}^{f_1 \times f_2 \times d} \to \mathbb{R}^d$ by g, such that for any $h = (h^1, \ldots, h^d) \in \mathbb{R}^{f_1 \times f_2 \times d}$,

$$\tilde{g}(h) := (g(h^1), \ldots, g(h^d)),$$

where h^i is the ith feature dimension projection of h for $i \in [d]$.

Now we are ready to give a rigorous definition of the pooling operator and pooling layers.

Definition 5.4 (Pooling Operator). Let $\mathcal{P} : \mathcal{F}([W] \times [H], \mathbb{R}^d) \to \mathcal{F}([\tilde{W}] \times [\tilde{H}], \mathbb{R}^d)$ be the pooling operator with stride $(s_1, s_2) \in \mathbb{Z}^2$ and kernel size $f = (f_1, f_2) \in \mathbb{Z}^2$, and activation function $g : \mathbb{R}^{f_1 \times f_2} \to \mathbb{R}$ if the following condition holds. Let $(\tilde{W}, \tilde{H})$ be defined by Equation (5.2). For any $h \in \mathcal{F}([W] \times [H], \mathbb{R}^d)$ and any $x = (x_1, x_2) \in [\tilde{W}] \times [\tilde{H}]$,

$$\mathcal{P}(h)(x) = \tilde{g}(h((x - 1) \circ s + [f])).$$

Definition 5.5 (Pooling Layer). Let $h^{(l)}$: $\mathcal{F}([W_0] \times [H_0], \mathbb{R}^{d_0})$ → $\mathcal{F}([W_l] \times [H_l], \mathbb{R}^{d_0})$ be the lth layer of the neural network. Let $\mathcal{P}$: $\mathcal{F}([W_{l-1}] \times [H_{l-1}], \mathbb{R}^{d_{l-1}}) \to \mathcal{F}([W_l] \times [H_l], \mathbb{R}^{d_l})$ be the pooling operator with stride $s_{\mathcal{P}} = (s_1, s_2)$, kernel size $f_{\mathcal{P}} = (f_1, f_2)$ and activation function $g : \mathbb{R}^{f_1 \times f_2} \to \mathbb{R}$, where

$$h^{(l)} = \mathcal{P}(h^{(l-1)}).$$

Then $h^{(l)}$ is a pooling layer of the neural network.

The commonly used activation functions in pooling layers include the max function and the average function, which are permutation invariant.

Remark 5.5. In the pooling layer, there are no trainable weights. It is used to reduce the spatial dimension while keeping the key information through the pooling operation.

5.3.3.3 *LeNet architecture*

Now we are ready to give a clear explanation of LeNet. Figure 5.17 represents the architecture of LeNet, which is constructed by adding the following layers one by one:

(1) input layer;
(2) convolutional layer with the ReLU activation function;
(3) pooling layer;
(4) convolutional layer with the ReLU activation function;
(5) pooling layer;
(6) two dense layers;
(7) output layer (dense layer with softmax activation function).

The reason for choosing the softmax activation function for the output layer is that the problem we are solving is a classification problem, where the output layer is used to estimate the conditional probability of the output being each class. Therefore we need to make sure that each neuron in the output layer is non-negative, and the summation of all the neurons should be equal to 1. Both conditions are satisfied by using the softmax function as the activation function. Note also that the output of the second pooling layer is image data of the 3D tensor type. To use this as the input of the dense layer, we need to flatten the image data into a (potentially long) vector before feeding it into the dense layer.

5.3.3.4 *Visualizing CNN*

In general, the more convolution steps we have, the more complicated are the features our network is able to learn to recognize. For example, in Figure 5.24, for handwritten digit classification, a CNN may learn to detect edges from raw pixels in the first layer and to identify simple shapes and higher-level features, such as strokes, from the lower layers to higher layers. Figures 5.23 and 5.24 are produced by visualizing the neurons of each layer of LeNet5 on the MNIST dataset. Readers who are interested in the details of implementations of CNNs are referred to the following website.[12]

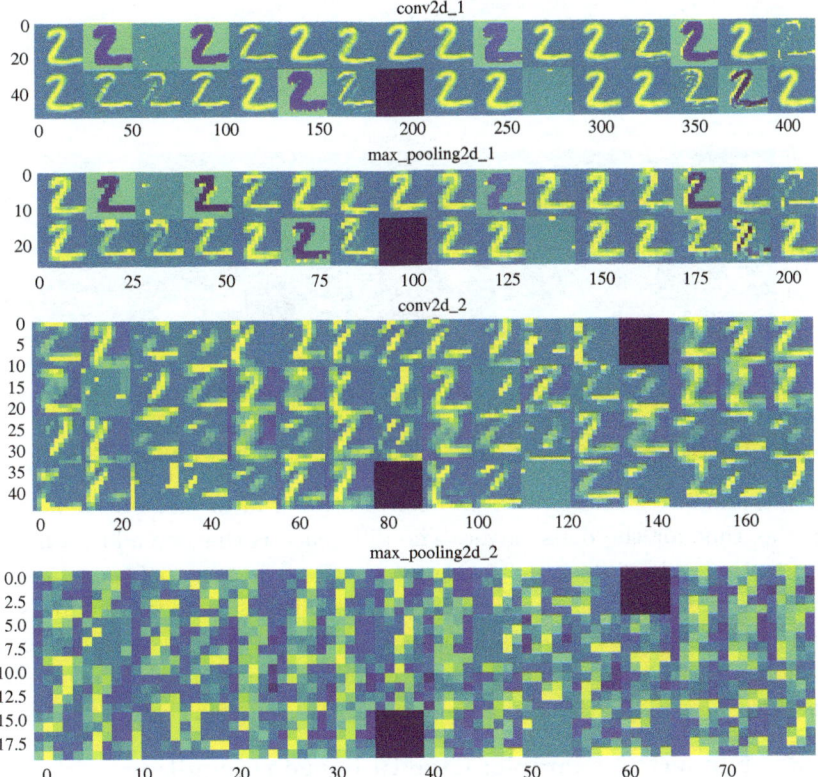

Figure 5.23. Visualizing a CNN trained on the handwritten digit 2 in the MNIST dataset.

---

[12]https://www.kaggle.com/amarjeet007/visualize-cnn-with-keras.

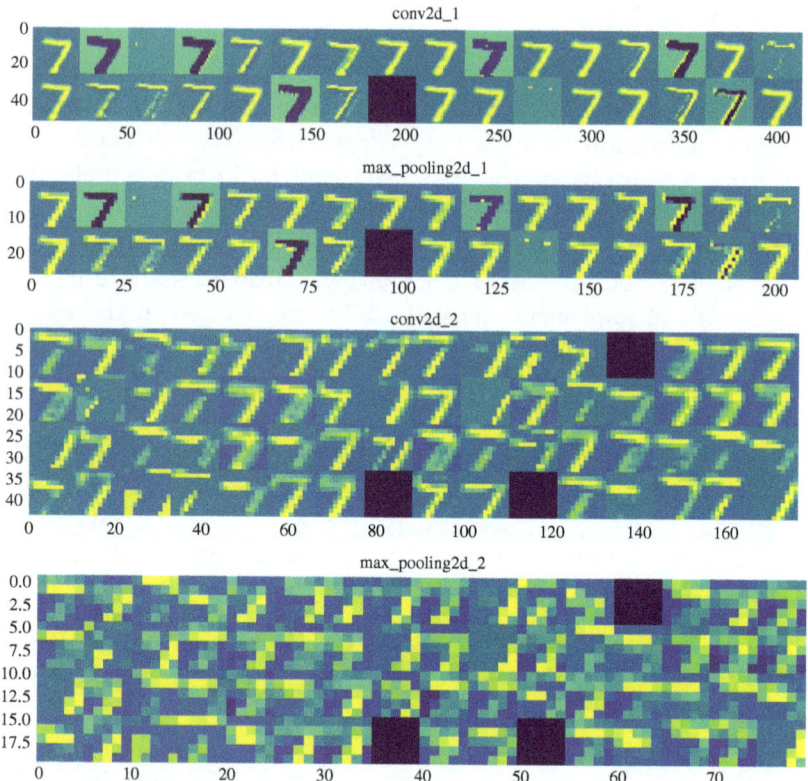

Figure 5.24. Visualizing a CNN trained on the handwritten digit 7 in the MNIST dataset.

5.3.4   Optimization

The parameter estimation for the convolutional layer is more or less the same as that for the dense layer. The difference is that we add an indicator function, which denotes the connectivity of neurons between two layers. This indicator function depends on the filter size. If two neurons are connected, then the weight is 1; otherwise the weight is 0. The backpropagation algorithm can still apply to compute the derivatives of the loss functions with respect to the parameters in the convolutional layer.

5.3.5   Numerical example: Cifar10 image recognition

The Cifar10[13] dataset is one of the most popular datasets for image recognition in machine learning research. The objective of the Cifar10 dataset

---

[13]https://www.cs.toronto.edu/~kriz/cifar.html.

is to design an algorithm for image classification. Image classification can be regarded as a supervised learning problem with the input–output pair (Input X, Output Y) $\sim$ (Image, Label).

Let us have a look at samples of the image classification task from the Cifar10 dataset. As its name suggests, there are ten different classes in Figure 5.25, and the corresponding label is given above the image.[14] The best results on Cifar10 data can achieve an accuracy as high as 96.53% using fractional max pooling [Graham (2014)]. DenseNet has achieved a state-of-the-art result comparable to fractional max pooling with a more compact architecture. [Huang *et al.* (2017)].

Listings 5.10 and 5.11 provide the Python codes to pre-process and implement LeNet for the Cifar10 data, respectively.

The summary of our LeNet model is given in Figure 5.26, which shows that our model has 1,204,682 trainable parameters. You may think that is a huge number of parameters. But if you were to construct a fully connected neural network with the same number of layers and hidden layers, the parameter number would much greater—more than 10^9!

Figures 5.27 and 5.28 show the loss function (cross entropy) and the accuracy of the LeNet model in both the training set and testing set with respect to the number of epochs. You can see that during the training stage,

Figure 5.25. Samples from Cifar10 dataset.

---

[14]http://rodrigob.github.io/are_we_there_yet/build/classification_datasets_res ults.html provides an excellent overview of standard datasets in image recognition and the accuracy results of various methods.

```
1   from keras.datasets import cifar10
2   from keras.utils import to_categorical
3   # Import Cifar10 dataset
4   (x_cifar10_train, y_cifar10_train), (x_cifar10_test, y_cifar10_test) =
    ↪   cifar10.load_data()
5   # Normalize the input data
6   x_train = x_cifar10_train.astype('float32')/255
7   x_test = x_cifar10_test.astype('float32')/255
8   # Convert class vectors to one-hot vector matrices
9   num_classes = 10
10  y_train = to_categorical(y_cifar10_train, num_classes)
11  y_test = to_categorical(y_cifar10_test, num_classes)
```

Listing 5.10. Pre-process Cifar10 dataset.

the CNN model is improving both in terms of the loss and the accuracy until the number of the epochs reaches 40. But it has already shown signs of overfitting as one may see that the accuracy gap between the training set and testing set is getting larger.

There are several ways to alleviate the overfitting problem, including

- data augmentation;
- dropout regularization;
- weight regularization.

5.3.5.1 *Data augmentation*

In image recognition, data augmentation is a commonly used trick. The main idea of data augmentation to help with the overfitting problem is to generate a lot of fake data using the original dataset to increase the robustness of the estimated model. It is evident that if we slightly rotate the image or slide the raw image to the left or right, the corresponding class label of this image should not change. Thus it is reasonable to increase the sample size by generating "fake" data by slightly rotating and shifting the image without a change of label. Following this idea, Listing 5.12 gives the code for data augmentation.

Figures 5.29 and 5.30 demonstrate that data augmentation helps resolve the overfitting issue, as there is no significant difference in the loss and accuracy between the train set and test set. It can be seen that by using the data augmentation, the accuracy of the testing set increases to around

```
1   from tensorflow.keras.models import Sequential
2   from tensorflow.keras.layers import Dense, Flatten, Conv2D, MaxPooling2D
3   from tensorflow.keras.optimizers import SGD
4   import time
5
6   # Build the first LeNet model
7   def Lenet_Model_BaseLine(n_hidden_neurons):
8       start_time = time.time()
9       print('Compiling Model ... ')
10      model = Sequential()
11      #Add 2 convolutional layers and pooling layers in turn
12      model.add(Conv2D(filters=n_hidden_neurons, kernel_size=(3, 3),
        ↪  input_shape=(32,32,3), activation ='relu'))
13      model.add(MaxPooling2D(pool_size=(2, 2)))
14      model.add(Conv2D(filters=n_hidden_neurons, kernel_size=(3,3),
        ↪  activation ='relu'))
15      model.add(MaxPooling2D(pool_size=(2, 2)))
16      # flatten the image data into a vector
17      model.add(Flatten())
18      model.add(Dense(512, activation ='relu'))
19      model.add(Dense(10, activation ='softmax'))
20      sgd = SGD(lr=0.001, decay=1e-6, momentum=0.9, nesterov=True)
21      model.compile(loss='categorical_crossentropy', optimizer=sgd,
        ↪  metrics=['accuracy'])
22      print('Model compield in {0} seconds'.format(time.time() -
        ↪  start_time))
23      return model
24
25  Lenet_Model = Lenet_Model_BaseLine(n_hidden_neurons=64)
26  print(Lenet_Model.summary())
27  hist_Lenet_Model= Lenet_Model.fit(x_train, y_train, epochs=40, batch_size
    ↪  = 128, validation_data=(x_test, y_test), verbose=1)
```

Listing 5.11. Implement CNN for classification using LeNet architecture.

75% in Figure 5.30, which is significantly higher than that of the baseline without data augmentation (69%, see Figure 5.28).

5.3.5.2 *Dropout regularization*

Dropout is one popular regularization method in the neural network, which was first proposed in [Srivastava *et al.* (2014)]. It randomly inactivates a proportion of neurons and learns a simplified model at each iteration. Due to the dropout, the network cannot rely heavily on a single weight as it may be eliminated. Therefore the learning process becomes more robust,

```
Compiling Model ...
Model compield in 0.12131261825561523 seconds
```

Layer (type)	Output Shape	Param #
conv2d_3 (Conv2D)	(None, 30, 30, 64)	1792
max_pooling2d_3 (MaxPooling2	(None, 15, 15, 64)	0
conv2d_4 (Conv2D)	(None, 13, 13, 64)	36928
max_pooling2d_4 (MaxPooling2	(None, 6, 6, 64)	0
flatten_2 (Flatten)	(None, 2304)	0
dense_3 (Dense)	(None, 512)	1180160
dense_4 (Dense)	(None, 10)	5130

```
Total params: 1,224,010
Trainable params: 1,224,010
Non-trainable params: 0
```

Figure 5.26. Summary of LeNet model in the Cifar10 example.

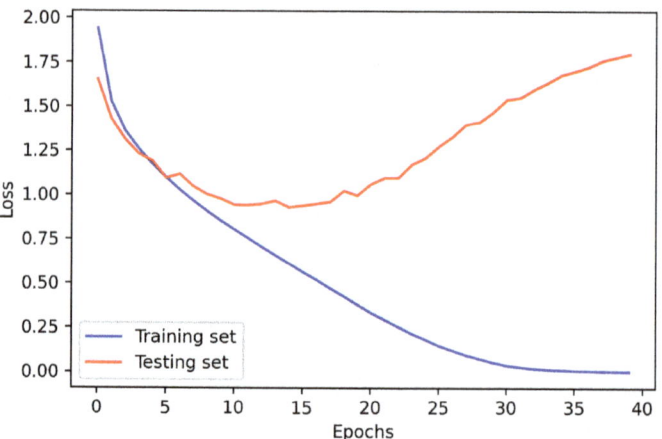

Figure 5.27. Loss of LeNet model in the training and testing set.

and dropout can avoid the overfitting issue effectively.[15] We give the implementation of a dropout method using Keras in Listing 5.13.

---

[15]The mathematical explanation behind dropout can be found in [Baldi and Sadowski (2013)].

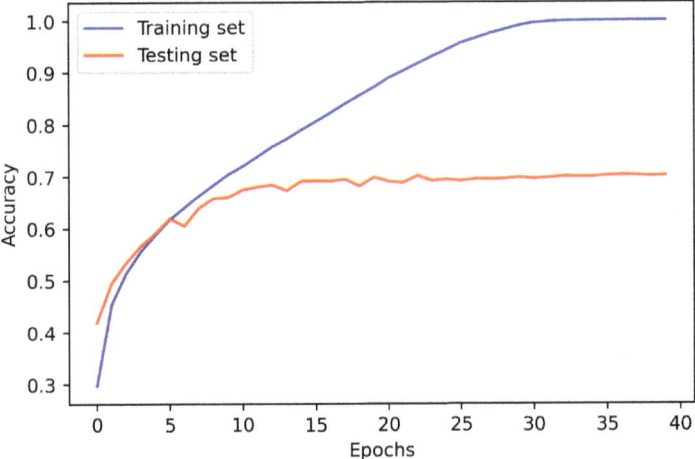

Figure 5.28. Accuracy of LeNet model in the training and testing set.

```
1   from keras.preprocessing.image import ImageDataGenerator
2
3   # Use data augmentation to avoid overfitting issue
4   iterations = 391
5   epochs = 160
6
7   datagen = ImageDataGenerator(horizontal_flip=True,
8               width_shift_range=0.125, height_shift_range=0.125,
                ↪  fill_mode='constant',cval=0)
9   datagen.fit(x_train)
10
11  # Start training
12  Lenet_da_Model  = Lenet_Model_BaseLine(n_hidden_neurons)
13  hist_Lenet_da = Lenet_da_Model.fit_generator(datagen.flow(x_train,
    ↪  y_train,batch_size=128), steps_per_epoch=iterations, epochs=epochs,
    ↪  validation_data=(x_test, y_test))
```

Listing 5.12. Use data augmentation to train LeNet.

The next question is how to choose the dropout rate. In general, for more complicated layers (more likely to be overfitting), we may need a higher dropout rate. For most applications, the hyper-parameter tuning is conducted by cross-validation. In Listing 5.14, we use the scikit-learn library to implement cross-validation to tune the parameter dropout rate to improve the accuracy of the LeNet model in the testing set. Figure 5.31

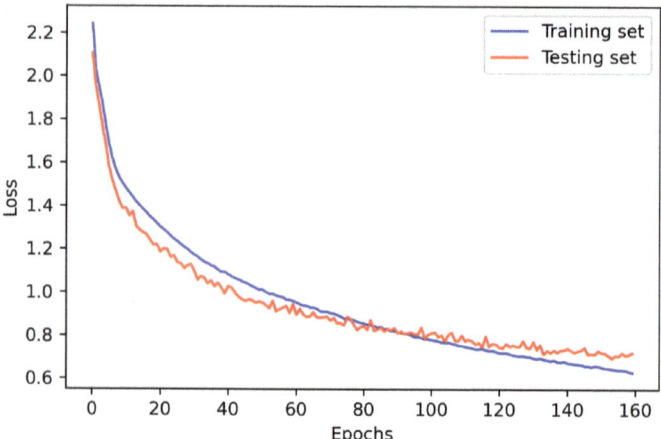

Figure 5.29. The loss of LeNet with data augmentation in the training and testing set.

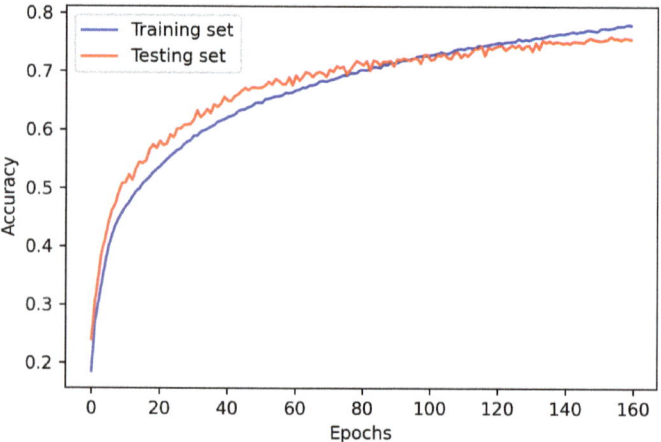

Figure 5.30. The accuracy of LeNet with data augmentation in the training and testing set.

suggests that the dropout rate should be set to 0.25 (average accuracy is 72.7%) among the three possible values $\{0.1, 0.25, 0.5\}$. One could make the possible numbers of the dropout rate larger to get a better dropout rate.

5.3.5.3 *Weight regularization*

As with Lasso and Ridge in the linear regression case, we can also add the l_1 or l_2 penalty term to the loss function in CNN models to avoid overfitting. To add the penalty term to the parameters in the CNN model, we simply

```
1   def Lenet_dp_Model(n_neurons, dropout_rate):
2       model = Sequential()
3       model.add(Conv2D(n_neurons, (3, 3), input_shape=(32,32, 3),activation
    ↪   = 'relu' ))
4       model.add(MaxPooling2D(pool_size=(2, 2)))
5       model.add(Dropout(dropout_rate))
6       model.add(Conv2D(64, (3,3), activation = 'relu'))
7       model.add(MaxPooling2D(pool_size=(2, 2)))
8       model.add(Dropout(dropout_rate))
9       model.add(Flatten())
10      model.add(Dense(512, activation = 'relu'))
11      model.add(Dropout(dropout_rate))
12      model.add(Dense(10, activation = 'softmax'))
13      sgd = SGD(lr=0.001, decay=1e-6, momentum=0.9, nesterov=True)
14      model.compile(loss='categorical_crossentropy', optimizer=sgd,
    ↪   metrics=['accuracy'])
15      return model
16
17  Ldp = Lenet_dp_Model(n_neurons =50, dropout_rate = 0.25)
18  hist_Ldp = Ldp.fit(x_train, y_train, epochs=800,  validation_data=(x_test,
    ↪   y_test), verbose=1)
```

Listing 5.13. Using dropout regularization to train LeNet.

```
Model compield in 0.08399224281311035 seconds
Best: 0.727060 using {'dropout_rate': 0.25}
0.720080 (0.005911) with: {'dropout_rate': 0.1}
0.727060 (0.008732) with: {'dropout_rate': 0.25}
0.634740 (0.008738) with: {'dropout_rate': 0.5}
```

Figure 5.31. Screenshot of the output of cross-validation for the dropout rate.

specify the "kernel_regularizer" when constructing the convolutional layer (see Listing 5.15).

5.4   Recurrent Neural Network

Up to now, we have discussed two main types of neural networks, ANN and CNN. The ANN is the base model in neural networks, while the CNN is highly efficient for image recognition. However, in both the ANN and CNN, neurons in the same layer are not directly connected. This type of network is called a *feed-forward* neural network.

This kind of feed-forward neural network may not always be the best model to describe the empirical data—sequential data in particular. In this section, we discuss the Recurrent Neural Network (RNN), which does not

```
1   from sklearn.model_selection import GridSearchCV
2   from keras.wrappers.scikit_learn import KerasClassifier
3
4   # create model
5   model = KerasClassifier(build_fn=Ldp, n_neurons=32,  epochs=160,
    ↪   batch_size=128, verbose=0)
6   # define the grid search parameters
7   dropout_rate = [0.1, 0.25, 0.5]
8   param_grid = dict(dropout_rate=dropout_rate)
9
10  grid = GridSearchCV(estimator=model, param_grid=param_grid)
11  grid_result = grid.fit(x_train, y_train)
12  # summarize results
13  print("Best: %f using %s" % (grid_result.best_score_,
    ↪   grid_result.best_params_))
14  means = grid_result.cv_results_['mean_test_score']
15  stds = grid_result.cv_results_['std_test_score']
16  params = grid_result.cv_results_['params']
17  for mean, stdev, param in zip(means, stds, params):
18      print("%f (%f) with: %r" % (mean, stdev, param))
```

Listing 5.14. Using cross-validation for parameter tuning of the dropout rate.

```
1   from keras import regularizers
2
3   model.add(Conv2D(32, (3, 3), input_shape=(32,32,3),activation = 'relu',
    ↪   kernel_regularizer=regularizers.l1(regRate)))
```

Listing 5.15. Adding l_1 penalty term to parameters of the convolutional layer.

belong to the feed-forward neural network. The hidden layer of RNN is interconnected and RNN has the recurrent structure.

RRN and its variants have proved particularly useful and delivered state-of-the-art results in diverse sequential data mining tasks, such as online handwritten text recognition, speech recognition [Graves *et al.* (2013)], and natural language processing [Wang and Jiang (2015)].

5.4.1   Introduction and motivation

The recurrent neural network (RNN) is a class of artificial neural network where connections between nodes form a directed graph along a temporal sequence. Figure 5.32 describes the unit structure of the RNN. The RNN

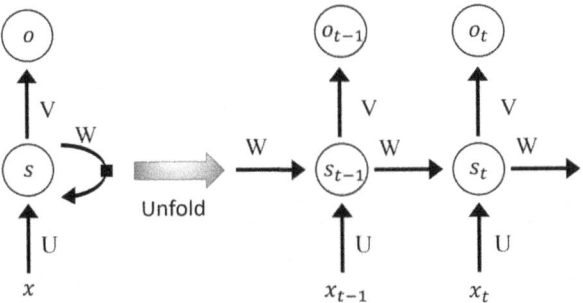

Figure 5.32. Unit structure of recurrent neural network.

provides an end-to-end training method to train a model for sequence labeling problems where the input–output alignment is unknown.

You may wonder why RNN is successful for analyzing sequential data. Let us explain the main intuition behind the RNN in the following example. Suppose that when you read the previous text "CNN", you naturally recall the content about CNNs in Section 5.3—that is a memory stored in your brain. In this case, we need a network where neurons have "memory" to capture temporal dependence, which can be well modeled by the recurrent structure.

To better understand the necessity for the recurrent structure, we use another example in natural language processing. Consider two sentences: "His answer is correct" and "They answer your question." The word "answer" appears in both sentences with different formations, as a pronoun and a verb. Suppose that we have a part-of-speech tagging task; the label of the word "answer" is not independent, as it has a higher probability of being a noun if it follows a pronoun. This shows how the previous unit is relevant to the current unit.

5.4.2 Sequential data

The input data of the recurrent neural network (RNN) is usually sequential data, e.g., stock price movements or video data.

On the abstract level, sequential data can be viewed as a function of time. In general, let us consider a mapping $F : \mathcal{T} \to E$, where $\mathcal{T} \subset \mathbb{R}^+$ is the set of possible time indexes. If $\mathcal{T}$ is a compact interval, the data are a continuous time series. If $\mathcal{T}$ is a finite set, F represents a discrete time series. In practice, most sequential datasets are discrete time series. In the following, let $\bar{x} := (x_t)_{t=1}^{T}$, where $x_t \in E$. In the example shown in

	Time	Type	Order ID	Size	Price	Direction
0	34200.017460	5	0	1	2238200	-1
1	34200.189608	1	11885113	21	2238100	1
2	34200.189608	1	3911376	20	2239600	-1
3	34200.189608	1	11534792	100	2237500	1
4	34200.189608	1	1365373	13	2240000	-1
5	34200.189608	1	11474176	2	2236500	1
6	34200.189608	1	1847685	100	2240000	-1
7	34200.189608	1	3920359	15	2236000	1
8	34200.189608	1	3578212	4	2240000	-1
9	34200.189608	1	4632045	100	2235000	1

Figure 5.33. Snapshot of the first nine messages of trades on Amazon (AMZN).

Figure 5.33, $\mathcal{T} = \{0, 1, 2, \ldots, 8\}$, and the data is a 6 dimensional time series of length $|\mathcal{T}|$; the number of possible time indexes is 9.

5.4.3 Model

RNN proposes a universal model for a continuous function on sequential data. The discrete RNN represented in Figure 5.32 is composed of three layers, i.e., the input layer, the hidden layer and the output layer, and it is defined as follows:

- Input Layer ($l^0 : \mathbb{R}^{T \times d} \to \mathbb{R}^{T \times d}$): $\forall \bar{x} \in \mathbb{R}^{T \times d}$,

$$\bar{x} \mapsto \bar{x}.$$

- Hidden Layer ($l^1 : \mathbb{R}^{T \times d} \to \mathbb{R}^{T \times n_1}$): $\forall \bar{x} \in \mathbb{R}^{T \times d}$,

$$\bar{x} \mapsto \bar{s} := (s_t)_{t=1}^T,$$

where $s_t = h(U x_t + W s_{t-1})$. Here U is an $n_1 \times d$ matrix of weights in front of x_t and W is an $n_1 \times n_1$ matrix of weights in front of s_{t-1}. x_t and s_t are column vectors. The multiplication is understood as matrix multiplication. h is the activation function of the hidden layers.

- Output Layer for sequential output ($l^2 : \mathbb{R}^{T \times d} \to \mathbb{R}^{T \times e}$) : $\forall \bar{x} \in \mathbb{R}^{T \times d}$,

$$\bar{x} \mapsto \bar{o} = (o_t)_{t=1}^T,$$

where $o_t = g(V s_t)$. Here g is the activation function of the output layer, and V is an $e \times n_1$ matrix of weights. In other words,

$$l^2(\bar{x}) = g(V l^1(\bar{x}^T)).$$

Table 5.4. Model of RNN.

- Input Layer $(l^0) : \bar{x} \mapsto \bar{x}$.
- Hidden Layer $(l^1) : \bar{x} \mapsto \bar{s}$,

$$s_t = h(Ux_t + Ws_{t-1}).$$

- Output Layer $(l^2) : \bar{x} \mapsto \bar{o} = (o_t)_{t=1}^T$ or o_T,

$$o_t = g(Vs_t).$$

The output layer for the vector valued output $(l^2 : \mathbb{R}^{T \times d} \to \mathbb{R}^e), \forall \bar{x} \in \mathbb{R}^{T \times d}$, is

$$\bar{x} \mapsto o_T. \tag{5.4}$$

Here (U, W, V) are RNN parameters, which can be trained from data. The output of the RNN can be a time series or a vector-valued variable. For example, the RNN can be used for sentence translation (time series) and classification of handwritten character trajectories (vector-valued output). Note that the hidden neurons have a recurrent structure, which means that s_t depends not only on the current input x_t, but also on s_{t-1}. The summary of the RNN model is given in Table 5.4.

5.4.4 Optimization: Backpropagation through time

Thanks to the recursive structure of the RNN, one can design an effective algorithm to compute the derivatives of the loss function with respect to the model parameters backwards. The direction of propagation is along time in RNN, not along the depth of layers as in ANN. Therefore, the algorithm is called *backpropagation through time* (BPTT).

For a task of predicting sequential output, the loss function is often in the additive form of the following equation:

$$\text{Loss Function:} \quad L(\bar{o}, \bar{y}) = \sum_{t=1}^{T} E_t(o_t, y_t),$$

where y_t and o_t are the actual/estimated output at time t, respectively: $\bar{o} = (o_t)_{t=1}^T$ and $\bar{y} = (y_t)_{t=1}^T$. The simplest choice for $E_t(o, y)$ is the squared

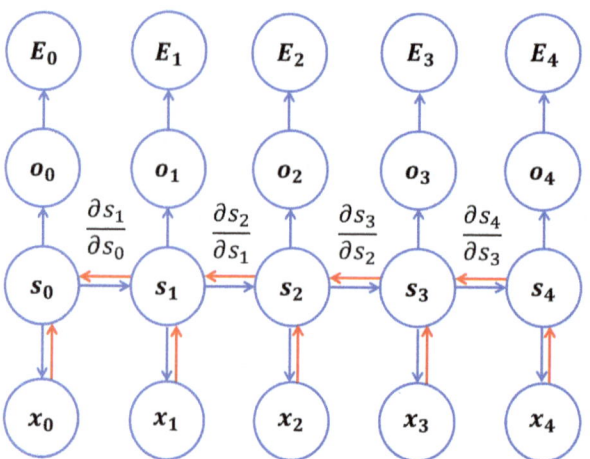

Figure 5.34. Illustration of backpropagation through time method.

error of o and y for the regression problem. However, E_t can be in a more complicated form, e.g., a time-dependent function.

To estimate the optimal parameters of the RNN, we use GD based numerical optimization methods. This requires computing the total differential $\frac{dE_t}{dV}$, $\frac{dE_t}{dU}$ and $\frac{dE_t}{dW}$. By the chain rule (Figure 5.34), it holds that

$$\frac{dE_t}{dV} = \frac{\partial E_t}{\partial o_t} \cdot \frac{\partial o_t}{\partial V}$$

$$\frac{dE_t}{dU} = \frac{\partial E_t}{\partial o_t} \cdot \frac{\partial o_t}{\partial U} = \frac{\partial E_t}{\partial o_t} \frac{\partial o_t}{\partial s_t} \cdot \frac{ds_t}{dU}$$

$$\frac{dE_t}{dW} = \frac{\partial E_t}{\partial o_t} \frac{\partial o_t}{\partial s_t} \cdot \frac{ds_t}{dW}$$

Thus the computation of $\frac{dE_t}{dV}$, $\frac{dE_t}{dW}$ and $\frac{\partial E_t}{\partial o_t}$ boils down to computing

- $\frac{\partial E_t}{\partial o_t}$, $\frac{\partial o_t}{\partial s_t}$, $\frac{\partial o_t}{\partial V}$;
- $\frac{ds_t}{dW}$, $\frac{ds_t}{dU}$.

First let us explain the derivation of the partial derivatives $\frac{\partial E_t}{\partial o_t}$, $\frac{\partial o_t}{\partial s_t}$, $\frac{\partial o_t}{\partial V}$. All these derivatives can be computed explicitly.

[a] *Compute* $\frac{\partial E_t}{\partial o_t}$. For example, if E_t is chosen to be mean squared error, then it holds that

$$\frac{\partial E_t}{\partial o_t} = \frac{\partial (o_t - y_t)^2}{\partial o_t} = 2(o_t - y_t).$$

[b] *Compute* $\frac{\partial o_t}{\partial s_t}$ *and* $\frac{\partial o_t}{\partial V}$. We use $\frac{\partial o_t}{\partial s_t}$ as an example to illustrate the derivations. Readers can follow the same method to compute the other partial derivatives $\frac{\partial o_t}{\partial V}$.

Recall that $o_t = g(V s_t)$ where $s_t \in \mathbb{R}^{n_1}$, $o_t \in \mathbb{R}^e$ and V is a matrix of size (e, n_1). $\frac{\partial o_t}{\partial s_t}$ is a matrix of size (e, n_1), i.e.,

$$\frac{\partial o_t}{\partial s_t} = \left(\frac{\partial o_t^i}{\partial s_t^j} \right)_{i \in [e], j \in [n_1]}. \tag{5.5}$$

Lemma 5.3. *If* $g : \mathbb{R} \to \mathbb{R}$ *is differentiable and* $o_t = g(V s_t)$, *then it holds that* $\forall i \in [e]$ *and* $j \in [n_1]$,

$$\frac{\partial o_t^i}{\partial s_t^j} = V_{ij} g' \left(\sum_{k=1}^{n_1} V_{ik} s_t^k \right). \tag{5.6}$$

Proof. For any $\forall i \in [e]$,

$$o_t^i = g((V s_t)^i = g \left(\sum_{k=1}^{n_1} V_{ik} s_t^k \right).$$

Then by the chain rule, it follows that for any $j \in [n_1]$,

$$\frac{\partial o_t^i}{\partial s_t^j} = V_{ij} g' \left(\sum_{k=1}^{n_1} V_{ik} s_t^k \right).$$

$\square$

In the following, we focus the discussion on the derivation of $\frac{ds_t}{dU}$ and $\frac{ds_t}{dW}$. By the definition of s_t, the recurrence of $\frac{ds_t}{dU}$ and $\frac{ds_t}{dW}$ holds as follows:

$$s_t = h(U x_t + W s_{t-1}) \implies \frac{ds_t}{dU} = \frac{\partial s_t}{\partial U} + \frac{\partial s_t}{\partial s_{t-1}} \frac{ds_{t-1}}{dU},$$

$$\frac{ds_t}{dW} = \frac{\partial s_t}{\partial W} + \frac{\partial s_t}{\partial s_{t-1}} \frac{ds_{t-1}}{dW}.$$

where $\frac{\partial s_t}{\partial W}$, $\frac{\partial s_t}{\partial s_{t-1}}$ and $\frac{\partial s_t}{\partial W}$ can be computed explicitly using a similar calculation shown in the proof of Lemma 5.3.

Thus by exploring the recurrent structure of $\frac{ds_t}{dU}$ and $\frac{ds_t}{dW}$, we have the formula for $\frac{ds_t}{dU}$ and $\frac{ds_t}{dW}$ in Lemma 5.4.

Lemma 5.4 (Recurrent Structure of $\frac{ds_t}{dU}$ and $\frac{ds_t}{dW}$). *For any* $t \in \{1, 2, \ldots, T\}$,

$$\frac{ds_t}{dU} = \frac{\partial s_t}{\partial U} + \sum_{k=0}^{t-1} \left(\prod_{j=k+1}^{t} \frac{\partial s_j}{\partial s_{j-1}} \right) \frac{\partial s_k}{\partial U}, \tag{5.7}$$

$$\frac{ds_t}{dW} = \frac{\partial s_t}{\partial W} + \sum_{k=0}^{t-1} \left(\prod_{j=k+1}^{t} \frac{\partial s_j}{\partial s_{j-1}} \right) \frac{\partial s_k}{\partial W}. \qquad (5.8)$$

Since $s_t = h(U x_t + W s_{t-1})$ and s_{t-1} also depends on U, it follows that:

$$\frac{ds_t}{dU} = \frac{\partial s_t}{\partial U} + \frac{\partial s_t}{\partial s_{t-1}} \frac{ds_{t-1}}{dU}.$$

Applying the above equation for the term $\frac{ds_{t-1}}{dU}$, it follows that

$$\begin{aligned}
\frac{ds_t}{dU} &= \frac{\partial s_t}{\partial U} + \frac{\partial s_t}{\partial s_{t-1}} \frac{ds_{t-1}}{dU} \\
&= \frac{\partial s_t}{\partial U} + \frac{\partial s_t}{\partial s_{t-1}} \left(\frac{\partial s_{t-1}}{\partial U} + \frac{\partial s_{t-1}}{\partial s_{t-2}} \frac{ds_{t-2}}{dU} \right).
\end{aligned}$$

Repeating this procedure until reaching $t = 0$, we have the formula Equation (5.7). The rigorous proof can be made as follows by induction.

Proof. The proof of the statement for $\frac{ds_t}{dW}$ is the same as that for $\frac{ds_t}{dU}$. Thus here we only present the proof for $\frac{ds_t}{dU}$—that is, Equation 5.7. For $t = 1$, Equation (5.7) is simplified to

$$\frac{ds_1}{dU} = \frac{\partial s_1}{\partial U} + \frac{\partial s_1}{\partial s_0} \frac{ds_0}{dU}.$$

This is true because of the chain rule.

Suppose that Equation (5.7) is true for $t - 1$. Let us check whether it is true for t. By the chain rule, it follows that

$$\frac{ds_t}{dU} = \frac{\partial s_t}{\partial U} + \frac{\partial s_t}{\partial s_{t-1}} \frac{ds_{t-1}}{dU}.$$

By induction assumption, it follows that

$$\begin{aligned}
\frac{ds_t}{dU} &= \frac{\partial s_t}{\partial U} + \frac{\partial s_t}{\partial s_{t-1}} \left(\frac{\partial s_{t-1}}{\partial U} + \sum_{k=0}^{t-2} \left(\prod_{j=k+1}^{t-1} \frac{\partial s_j}{\partial s_{j-1}} \right) \frac{\partial s_k}{\partial U} \right) \\
&= \frac{\partial s_t}{\partial U} + \frac{\partial s_t}{\partial s_{t-1}} \frac{\partial s_{t-1}}{\partial U} + \frac{\partial s_t}{\partial s_{t-1}} \sum_{k=0}^{t-2} \left(\prod_{j=k+1}^{t-1} \frac{\partial s_j}{\partial s_{j-1}} \frac{\partial s_k}{\partial U} \right)
\end{aligned}$$

$$= \frac{\partial s_t}{\partial U} + \frac{\partial s_t}{\partial s_{t-1}} \frac{\partial s_{t-1}}{\partial U} + \sum_{k=0}^{t-2} \left(\prod_{j=k+1}^{t} \frac{\partial s_j}{\partial s_{j-1}} \right) \frac{\partial s_k}{\partial U}$$

$$= \frac{\partial s_t}{\partial U} + \sum_{k=0}^{t-1} \left(\prod_{j=k+1}^{t} \frac{\partial s_j}{\partial s_{j-1}} \right) \frac{\partial s_k}{\partial U} (\Leftarrow x_{1:t}).$$

Thus Equation (5.7) is true for t. The proof is complete by induction. $\square$

5.4.5 Limitation of RNN

According to Lemma 5.4, $\frac{ds_t}{dU}$ depends on $(x_{1:t})$, which results in the heavy computational cost. Besides this, it may lead to vanishing/exploding gradient problems. This can be mitigated by *truncated backpropagation through time* (TBPTT), see Equation (5.9).

$$\frac{ds_t}{dU} = \frac{\partial s_t}{\partial U} + \sum_{k=0}^{t-1} \left(\prod_{j=k+1}^{t} \frac{\partial s_j}{\partial s_{j-1}} \right) \frac{\partial s_k}{\partial U} (\Leftarrow x_{1:t}) \tag{5.9}$$

$$\approx \frac{\partial s_t}{\partial U} + \sum_{k=t-K}^{t-1} \left(\prod_{j=k+1}^{t} \frac{\partial s_j}{\partial s_{j-1}} \right) \frac{\partial s_k}{\partial U} (\Leftarrow x_{t-K:t}). \tag{5.10}$$

TBPTT is a modified version of the BPTT training algorithm for recurrent neural networks and is depicted in Figure 5.35, where the sequence is processed one timestep at a time and periodically (k_1 timesteps), and the BPTT update is performed back for a fixed number of timesteps (k_2).

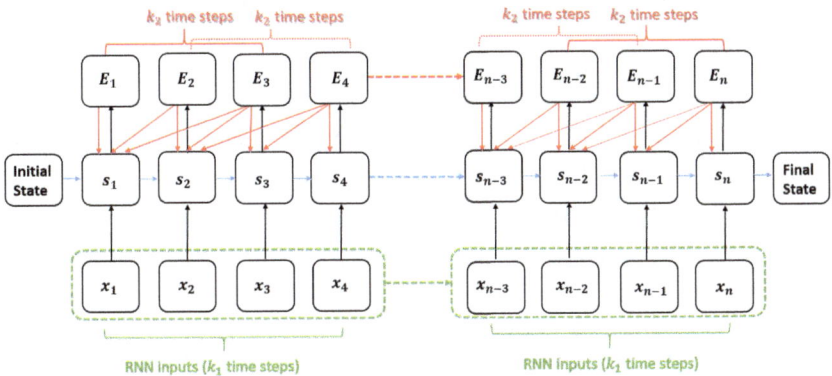

Figure 5.35. Illustration of TBPTT algorithm.

The TBPTT algorithm is summarized as follows:[16]

(1) Feed a sequence of k_1 timesteps of input and output pairs to the network.
(2) Unroll the network, then calculate the accumulated errors across k_2 timesteps.
(3) Roll-up the network and update model weights.
(4) Repeat above process.

Generally speaking, k_1 affects the duration of the training, giving how often weight updates are performed. k_2 should be large enough to capture the temporal structure in the problem for the network to learn. However, too high a value k_2 results in vanishing gradients. Here we present the architecture of two commonly used TBPTT algorithms, illustrated in Figure 5.36. For readers who are interested in TBPTT, please refer to the following material.[17]

Furthermore, vanishing/exploding gradient problems can also be relieved by introducing the *gate* mechanism. We discuss two variants of RNN in the next subsection. A summary of the limitations of RNN is given in Table 5.5.

5.4.6   Variants of RNN: LSTM and GRU

To relieve the vanishing and exploding gradient issues of RNN, several variants of the RNN model have been proposed, including long short-term memory (LSTM) [Hochreiter and Schmidhuber (1997)] and gated recurrent units (GRU) [Cho *et al.* (2014)].

5.4.6.1   *LSTM*

Let us first discuss LSTM. As we can see in Figure 5.37, a standard LSTM unit is composed of a cell, an input gate, an output gate and a forget gate. The cell remembers values over arbitrary time intervals, and the three gates regulate the flow of information into and out of the cell.

The mathematical formulation of an LSTM unit is given as follows [Hochreiter and Schmidhuber (1997)]:

$$f_t = \sigma_g(W_f x_t + U_f h_{t-1} + b_f)$$
$$i_t = \sigma_g(W_i x_t + U_i h_{t-1} + b_i)$$

---

[16]https://machinelearningmastery.com/gentle-introduction-backpropagation-ti me/.
[17]https://r2rt.com/styles-of-truncated-backpropagation.html.

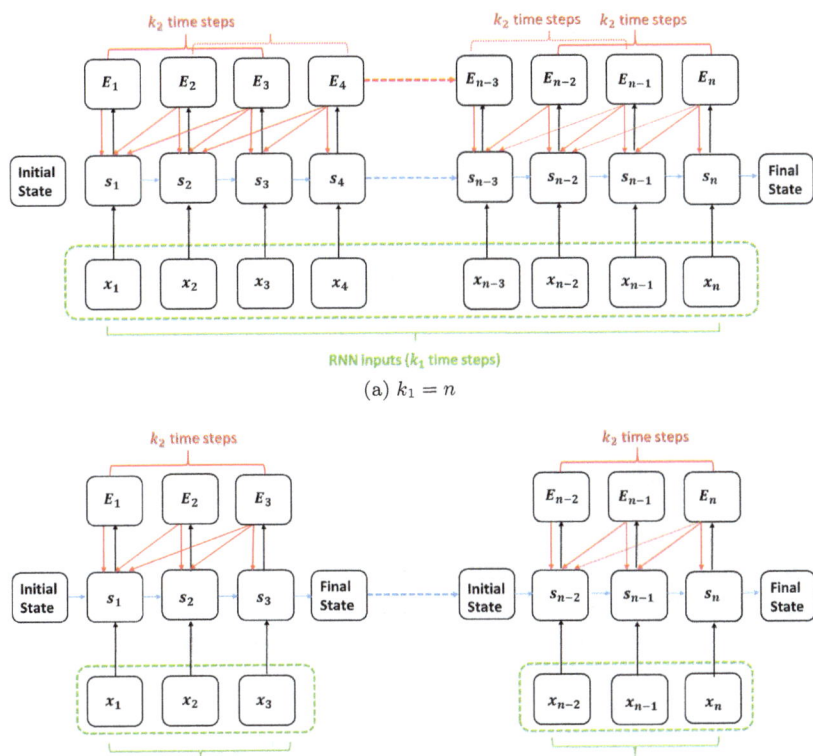

(a) $k_1 = n$

(b) $k_1 = k_2$. Tensorflow uses this TBPTT algorithm as default for its implementation

Figure 5.36. Different types of truncated backpropagation through time.

Table 5.5. Summary of the limitations of RNN.

The limitations of RNN are as follows:

- Problem: Heavy computational cost.

  Solution: Truncated backpropagation through time (TBPTT).

- Problem: Vanishing/exploding gradient problems.

  Solution: Long short term memory (LSTM); gated recurrent units (GRU).

$$o_t = \sigma_g(W_o x_t + U_o h_{t-1} + b_o)$$
$$c_t = f_t \circ c_{t-1} + i_t \circ \sigma_c(W_c x_t + U_c h_{t-1} + b_c)$$
$$h_t = o_t \circ \sigma_h(c_t),$$

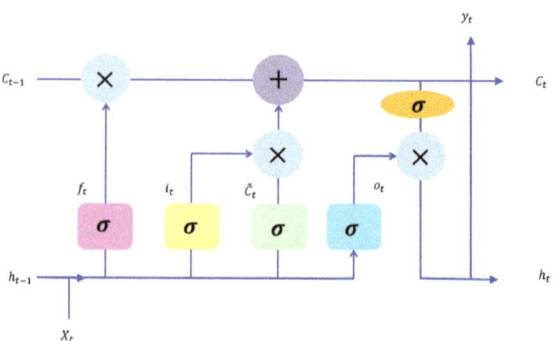

Figure 5.37. Long short-term memory.

where the initial values are $c_0 = h_0 = 0$, and the operator $\circ$ denotes the Hadamard product (elementwise product) and

- $x_t \in \mathbb{R}^d$: input vector of a LSTM unit;
- $f_t \in \mathbb{R}^h$: forget gate's activation vector;
- $i_t \in \mathbb{R}^h$: output gate's activation vector;
- $o_t \in \mathbb{R}^h$: output vector, also known as hidden state vector, of a LSTM unit;
- $c_t \in \mathbb{R}^h$: cell state vector;
- $W \in \mathbb{R}^{h \times d}, U \in \mathbb{R}^{h \times h}$ and $b \in \mathbb{R}^h$: weight matrices and bias vector parameters that need to be learned during training, where the superscripts d and h refer to the number of input features and number of hidden units, respectively.

Typically the activation functions are chosen as follows:

- σ_g is a sigmoid function;
- σ_c is hyperbolic tangent function;
- σ_h is a hyperbolic tangent function.

The motivation of LSTM came from the long-term dependency problem of the RNN. LSTM can successfully alleviate this problem by adding the cell state, which includes long-term information, at each time. LSTM has the ability to remove or add information to the cell state, carefully controlled by the gates. In LSTM, the network divides the state into the short-term state and the long-term state through the gating mechanism. Specifically, an LSTM cell can learn to recognize part of the input with long-lasting effects, store it in the long-term state, preserve it, and extract it whenever it is needed. At the same time, a forget gate can learn what redundant information to remove from the cell state. This network design explains why they have been amazingly successful at capturing long-term patterns in time series, long texts and more.

5.4.6.2 *GRU*

The GRU network is a variant of the RNN using a gating mechanism. Compared with LSTM, GRU has fewer parameters without an output gate. Therefore, GRU has been shown to exhibit better performance on smaller datasets than LSTM. However, it shown that GRU performs similarly to LSTM on larger datasets, e.g., polyphonic music modeling and speech signal modeling.

5.4.7 Numerical example: High frequency financial data prediction

In this subsection, following recent work [Sirignano and Cont (2018)], we explain how to use LSTM to forecast the future price direction of high-frequency financial data. Let us start with an example of a real-world dataset from LOBSTER[18]—an online platform to provide easy-to-use, high-quality limit order book data.[19]

Typically LOBSTER provides two types of data, i.e., a "message" and "orderbook" file for each active trading day of a selected ticker. The "orderbook" file contains the evolution of the limit order book up to the requested number of levels. The "message" file contains indicators for the type of event causing an update of the limit order book in the requested price range.[20]

When it comes to the sequential data, Figure 5.33 in Section 5.4.2 provides a snapshot of the "message" data, which includes the following fields:

- Time: Seconds after midnight with decimal precision of at least milliseconds and up to nanoseconds depending on the requested period.
- Type:
  (1) Submission of a new limit order;
  (2) Cancellation (partial deletion of a limit order);
  (3) Deletion (total deletion of a limit order);
  (4) Execution of a visible limit order;

---

[18]LOBSTER stands for Limit Order Book System—The Efficient Reconstructor. [Huang and Polak (2011)].
[19]https://lobsterdata.com/info/WhatIsLOBSTER.php.
[20]https://lobsterdata.com/info/DataStructure.php.

(5) Execution of a hidden limit order;
(6) Trading halt indicator.

- Order ID: Unique order reference number (assigned in order flow).
- Size: Number of shares.
- Price: Dollar price times 10,000 (i.e., A stock price of \$91.14 is given by 911,400).
- Direction:

    −1: Sell limit order;
    1: Buy limit order.

For example, the first row in Figure 5.33 indicates that at the time 34200.017460 seconds after midnight on June 21, 2012 there was a hidden sell limit order of size 1 to be executed at the price level \$223.82.[21]

Let $P_l^a(t)$ and $P_l^b(t)$ denote the ask and bid prices, respectively, of the lth level at time t. Similarly, $V_l^a(t)$ and $P_l^b(t)$ denote the ask and bid volume, respectively, of the lth level at time t. Let LOB_t denote the limit order book at time t, which includes the bid and ask prices of levels $1, 2, \ldots, 10$ and the corresponding volumes, i.e., $LOB_t = (P_l^a(t), V_l^a(t), P_l^b(t), V_l^b(t))_{l=1}^{10}$.

Figure 5.39a shows an example of a limit order book up to level 10 at a time. Here the x-axis represents volume level and the y-axis represents the price levels of the limit order book, where red and green colors represent bid and ask orders, respectively. For example, Figure 5.38 shows that the evolution of the execution number and trade volume of Amazon stock aggregated into every five minute interval in Figure 5.39a, the best bid is at the price of \$220.94 with the quantity 100 shares.

Figure 5.40 shows the first five orders in the limit order book. A short piece of python code, which uses the Numpy and Pandas libraries to import the message and orderbook data, is listed in Listing 5.16.

The mid-price is defined to be the average of the best bid and ask price (level 1) denoted by M_t, i.e.,

$$M_t := \frac{1}{2}(P_1^a(t) + P_1^b(t)).$$

Let τ_i denote the time when the ith change in the mid-price occurs. Thus at any τ_i, the mid-price is either going up or down. Let Y_i denote the sign of the next price movement,

$$Y_i := \mathbf{1}(M_{\tau_{i+1}} - M_{\tau_i} > 0).$$

---

[21]Figures 5.38 and 5.39a are produced by the demo code in `https://lobsterdata.com/info/help_codeHelp.php`.

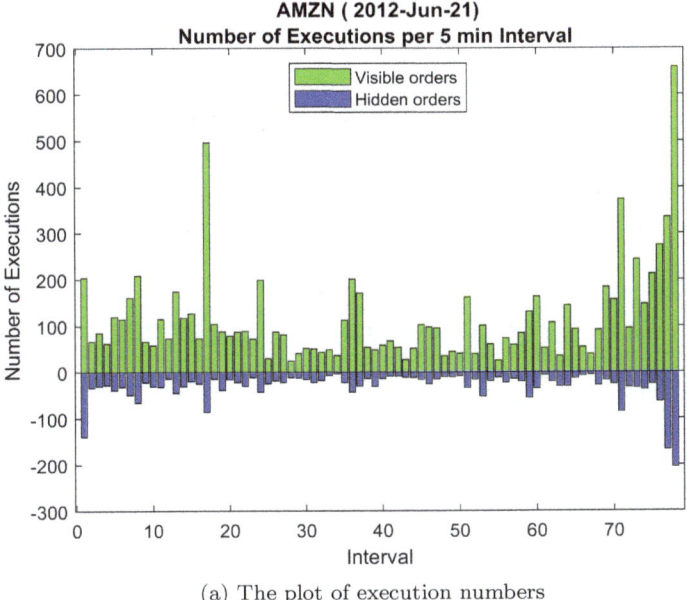

(a) The plot of execution numbers

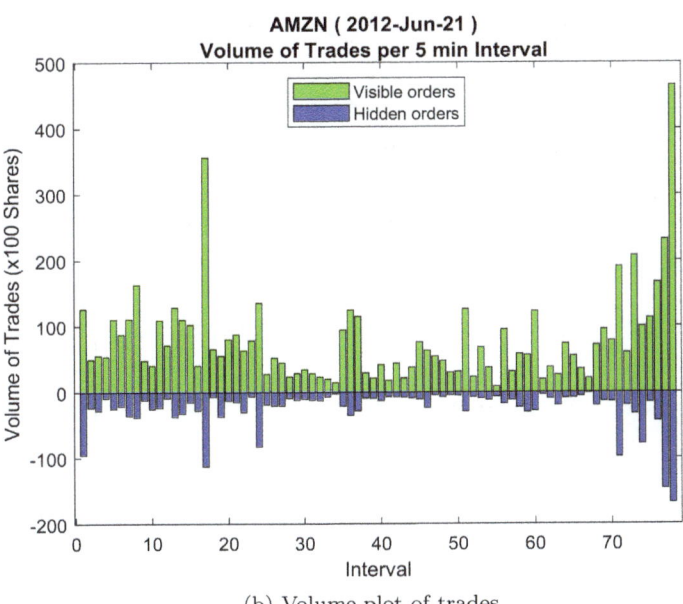

(b) Volume plot of trades

Figure 5.38. Plot of execution number and volume.

Figure 5.41 shows the first few entries of the orders that precede the next price change in the dataset. In Listing 5.17, we remove the events that do not have mid price change.

The input factor is the p-lagged values of LOB_{τ_i}, i.e.,

$$X_i := (LOB_{\tau_i}, LOB_{\tau_{i-1}}, \ldots, LOB_{\tau_{i-p+1}}).$$

Therefore the prediction of future price movement can be viewed as a binary classification problem with input and output pairs $(X_i, Y_i)_{i=1}^N$.

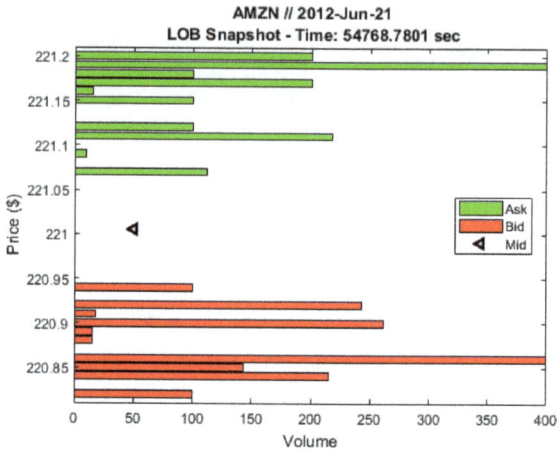

(a) A snapshot of LOB of AMZN on January 21 2013

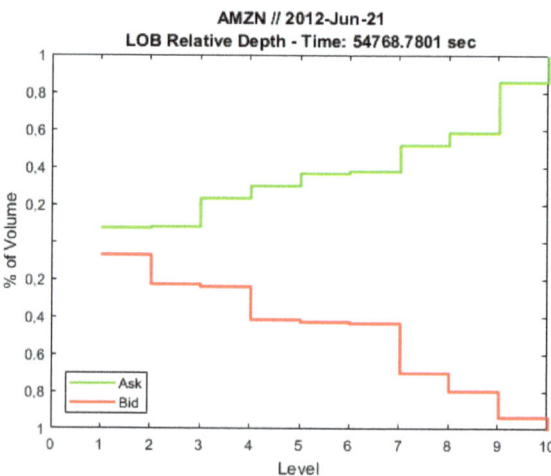

(b) A snapshot of the relative depth of AMZN on January 21 2013

Figure 5.39. Visualization of limit order book (LOB).

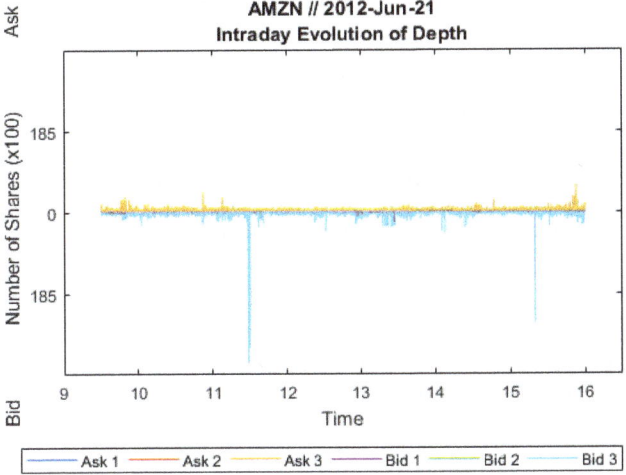

(c) Different color indicates the price of bid/ask at different levels

Figure 5.39. (*Continued*)

	ask price 1	ask size 1	bid price 1	bid size 1	ask price 2	ask size 2	bid price 2	bid size 2	...	ask price 10	ask size 10	bid price 10	bid size 10
0	223.95	100	223.18	100	223.99	100	223.07	200	...	229.8	100	218.97	100
1	223.95	100	223.81	21	223.99	100	223.18	100	...	229.8	100	220.2	100
2	223.95	100	223.81	21	223.96	20	223.18	100	...	229.43	100	220.2	100
3	223.95	100	223.81	21	223.96	20	223.75	100	...	229.43	100	220.25	5000
4	223.95	100	223.81	21	223.96	20	223.75	100	...	229.43	100	220.25	5000

Figure 5.40. The first five orders in the limit order book.

Data preparation using the LOBSTER data is provided in Listing 5.17. In this example, we use the LOB information of AAPL from 2014-01-02 to 2014-01-15 and split the train, validation and test set based on the ratios 0.8 : 0.1 : 0.1. Listing 5.18 prepares the LOB data ready for the LSTM model to train.

Before applying the LSTM model for the prediction task, we need to split the data into the train and test set, and normalize the data in the data pre-process procedure. The relevant codes are given in Listing 5.18.

The implementation of the single LSTM model for prediction of the next price movement is given in Listing 5.19. Figures 5.42 and 5.43 visualize the evolution of the loss function and the accuracy of the single-layer LSTM over a number of epochs.

We also implement a three-layer LSTM following [Sirignano and Cont (2018)] (Listing 5.20) and apply it to our dataset (replace lstm() with lstm3() on line 19 in Listing 5.19). One can find the corresponding accuracy

```
 1   # import the required libraries
 2   import numpy as np
 3   import pandas as pd
 4
 5   # import data
 6   def import_orderbook(filedir):
 7     ob = pd.read_csv(filedir,header=None).values.astype(float)
 8     ob[:,0::2]=(ob[:,0::2]/1e4) # normalization.
 9     df_ob =  pd.DataFrame(ob, columns = str_sets)
10     return ob, df_ob
11
12   def import_message(filedir):
13     message=pd.read_csv(filedir,header=None).values
14     df_message = pd.DataFrame(message, columns=['Time', 'Type', 'Order ID',
         ↪  'Size', 'Price', 'Direction'])
15     df_message[['Type', 'Order ID', 'Size', 'Price', 'Direction']] =
         ↪  df_message[['Type', 'Order ID', 'Size', 'Price',
         ↪  'Direction']].astype('int')
16     return message, df_message
17
18   orderbook, df_orderbook = import_orderbook('orderbook.csv')
19   message, df_message = import_message('message.csv')
```

Listing 5.16. Import limit order book data using Numpy and Pandas libraries.

	ask price 1	ask size 1	bid price 1	bid size 1	ask price 2	ask size 2	bid price 2	bid size 2	ask price 3	ask size 3	⋯	bid price 8	bid size 8	ask price 9	ask size 9	bid price 9	bid size 9	ask price 10	ask size 10	bid price 10	bid size 10
0	223.95	100.0	223.18	100.0	223.99	100.0	223.07	200.0	224.00	220.0	⋯	220.25	5000.0	229.43	100.0	220.20	100.0	229.80	100.0	218.97	100.0
31	223.95	100.0	223.81	21.0	223.96	20.0	223.75	100.0	223.99	100.0	⋯	223.07	200.0	224.49	100.0	223.04	100.0	224.50	5.0	223.00	10.0
43	223.95	100.0	223.75	74.0	223.96	306.0	223.65	2.0	223.99	100.0	⋯	222.62	100.0	224.49	100.0	221.30	4000.0	224.50	5.0	220.40	100.0
45	223.96	286.0	223.75	74.0	223.99	100.0	223.65	2.0	224.00	1451.0	⋯	222.62	100.0	224.50	5.0	221.30	4000.0	224.89	100.0	220.40	100.0
47	223.99	100.0	223.75	74.0	224.00	1451.0	223.65	2.0	224.24	20.0	⋯	223.00	10.0	224.89	100.0	222.62	100.0	226.77	100.0	221.30	4000.0

Figure 5.41. The collection of orders that are followed by the next price change.

and loss evolution of the three-layer LSTM w.r.t. the number of epochs in Figures 5.44 and 5.45, respectively. The three-layer LSTM model results in slightly better accuracy on the test set compared with that of the one-layer LSTM (Table 5.6). However, one may draw different conclusion of the performance comparison between those two models using a small dataset, e.g., one-day limit order book data.

The accuracy of the next price direction prediction we obtained here is significantly worse than that of [Sirignano and Cont (2018)]. In [Sirignano and Cont (2018)], an average accuracy of 70% is achieved in the out-of-sample data.

```
1    # Import library
2    import numpy as np
3    import pandas as pd
4
5    # Remove the events that do not have mid price change
6    def RemoveNoNextPriceChangeEvent(df_orderbook):
7      orderbook=df_orderbook.values
8      midprice=(orderbook[:,0]+orderbook[:,2])/2
9      sign_vec = np.sign(midprice[1:]-midprice[:-1])
10     index0 = np.where(sign_vec!=0)
11     orderbook = orderbook[:-1, :]
12     sign_vec = sign_vec[index0[0]]*0.5+0.5
13     sign_vec = sign_vec.astype('int')
14     label = keras.utils.to_categorical(sign_vec,
       ↪ num_classes=2,dtype='float32')
15     df_orderbook = df_orderbook.iloc[index0]
16     return df_orderbook.values, df_orderbook, label
17
18   orderbook, df_orderbook, label =
     ↪ RemoveNoNextPriceChangeEvent(df_orderbook)
19
20   # Prepare the data for the learning algorithm
21   def P_Lagged_data(X,p):
22     n=X.shape[0]
23     LaggedX=np.zeros([n-p,p,X.shape[1]])
24     for i in range(n-p):
25       LaggedX[i,:,:]=X[i:(i+p),:]
26     return(LaggedX)
27
28   def GenerateLaggedValueInputOutput(df_orderbook, label, p):
29     orderbook = df_orderbook.values
30     inputX = P_Lagged_data(orderbook,p)
31     outputY = label[p:]
32     return inputX, outputY
33
34   nTimes = df_orderbook.shape[0]
35   p = 50
36   inputX, outputY = GenerateLaggedValueInputOutput(df_orderbook, label, p)
```

Listing 5.17. Preparation of input–output data from LOBSTER dataset.

One may wonder why we use the same neural network and do not replicate the results in [Sirignano and Cont (2018)]. The reason is the different sample size of the dataset. In [Sirignano and Cont (2018)], the LOB data of 500 stocks over one year are pooled together to train an accurate and

```
1   # import libraries
2   from sklearn.model_selection import train_test_split
3   # Split the train and test set
4   # We set the parameter shuffle flag False, as we want
5   # to use the last 10\% of data for testing. The default
6   # flag (True) means the testing data is randomly selected.
7   X_train, X_test, y_train, y_test = train_test_split(inputX, outputY,
    ↪   test_size=0.1, shuffle = False)
8
9   # Data normalization
10  def Normalize3D_tensor_Data(X_train, X_test):
11    dim_train = np.shape(X_train)
12    X_train_s = np.zeros(dim_train, dtype = float)
13    X_test_s = np.zeros(np.shape(X_test), dtype = float)
14    for i in range(dim_train[1]):
15      for j in range(dim_train[2]):
16        scaler = preprocessing.StandardScaler().fit(X_train[:, i, :])
17        X_train_s[:, i, :] = scaler.transform(X_train[:, i, :])
18        X_test_s[:, i, :] = scaler.transform(X_test[:, i, :])
19      return X_train_s, X_test_s
20
21  X_train_scaled, X_test_scaled = Normalize3D_tensor_Data(X_train, X_test)
```

Listing 5.18. Prepare the limit order book data for the LSTM model.

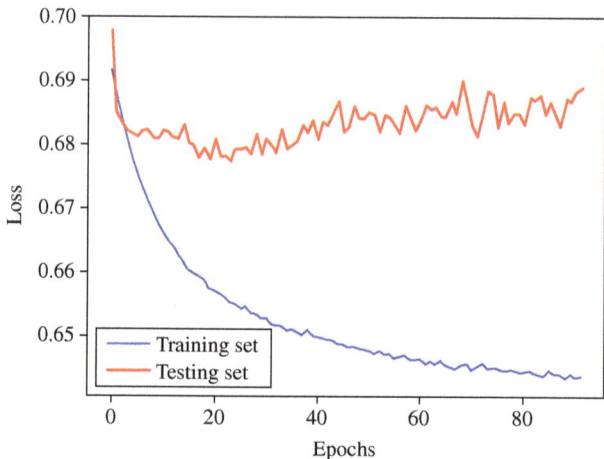

Figure 5.42. The loss of a single-layer LSTM w.r.t. the number of epochs in the training and testing sets.

```
1    # Import library
2    import tensorflow as tf
3    import keras
4    from tensorflow.keras.layers import Dense, LSTM, Dropout, Activation
5    from tensorflow.keras import optimizers,metrics
6    from tensorflow.keras.models import Sequential
7
8    def lstm(input_shape, nodes, dropout):
9        model=Sequential()
10       model.add(LSTM(nodes,dropout=dropout,
11       input_shape=input_shape, use_bias= True))
12       model.add(Dense(10, activation='relu', use_bias= True))
13       model.add(Dense(2,activation='softmax', use_bias= True))
14       adam=keras.optimizers.Adam(lr=0.01, beta_1=0.9, beta_2=0.999,
         ↪   epsilon=None, decay=0.0, amsgrad=False)
15       model.compile(loss='categorical_crossentropy', optimizer='adam',
16       metrics=[metrics.categorical_accuracy])
17       return model
18
19   model1=lstm(X_train_scaled.shape[1:], 30,0.15)
20   hist1 = model1.fit(X_train_scaled , y_train,validation_split=0.1,
       ↪   batch_size=2000, epochs=50, shuffle=True,verbose=1)
21   scores = model1.evaluate(X_test_scaled, y_test, verbose=1)
```

Listing 5.19. One-layer LSTM for prediction of the next price movement direction.

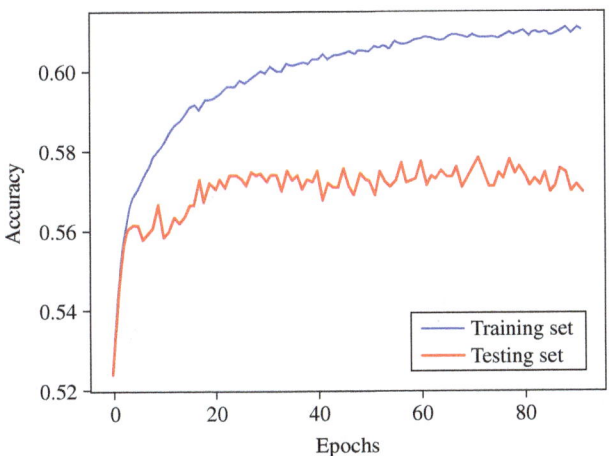

Figure 5.43. The accuracy of a single-layer LSTM w.r.t. the number of epochs in the training and testing sets.

Table 5.6. The performance of the single-layer LSTM model and the three-layer LSTM model for predicting the next price direction in the test data.

	Accuracy	Precision	Recall	f1 score
One-layer LSTM	56.32%	54.12%	76.46%	62.11%
Three-layer LSTM	57.57%	55.24%	75.27%	62.64%

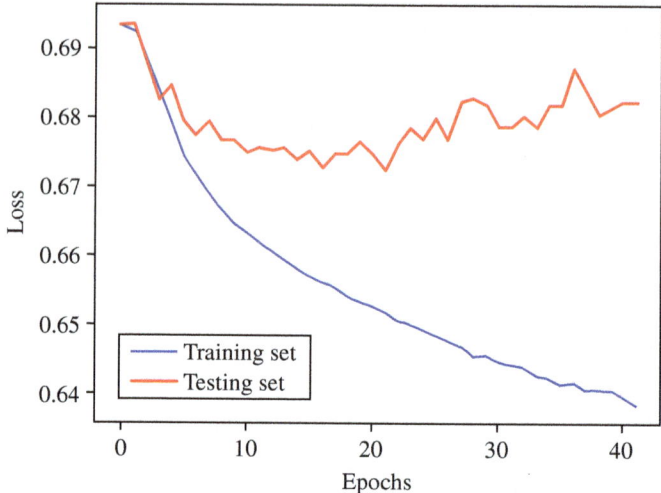

Figure 5.44. The loss of a three-layer LSTM w.r.t. the number of epochs in the training and testing sets.

robust forecasting model, while in our toy example, we use a single asset over 10 days. It is well known that deep learning techniques are data greedy. Without a large sample size, the application of deep learning risks poor predictive performance due to the overfitting issue. To verify this point, we vary the length of the training set, i.e., 1, 2, 4, 8, 16 and 32 days, and see how the out-of-sample accuracy changes w.r.t. the number of trading days in the training set. Figure 5.46 shows that as the sample size increases, the out-of-sample accuracy increases accordingly.[22]

---

[22]The sensitivity analysis w.r.t. data sample size is attributed to Hang Lou's work.

```
1   def lstm3(input_shape, nodes,rcd,d):
2       model=Sequential()
3       model.add(LSTM(nodes,return_sequences=True,
4       recurrent_dropout=rcd,dropout=d,
5       input_shape=input_shape, use_bias= True))
6       model.add(LSTM(nodes,return_sequences=True,
7       recurrent_dropout=rcd, dropout=d,  use_bias= True))
8       model.add(LSTM(nodes,recurrent_dropout=rcd,
9       dropout=d, use_bias= True))
10      model.add(Dense(50, activation='relu', use_bias= True))
11      model.add(Dense(2,activation='softmax'))
12      adam=keras.optimizers.Adam(lr=0.01, beta_1=0.9, beta_2=0.999,
        ↪  epsilon=None, decay=0.0, amsgrad=False)
13      model.compile(loss='categorical_crossentropy', optimizer='adam',
14      metrics=[metrics.categorical_accuracy])
15      return model
16
17  model2=lstm3(X_train_scaled.shape[1:],30, 0.25,0.2)
    ↪  print(model2.summary()) hist2
18  = model2.fit(X_train_scaled, y_train,validation_split=0.1, batch_size=500,
19  epochs=100, shuffle=True,verbose=1) scores2 =
    ↪  model2.evaluate(X_test_scaled,
20  y_test, verbose=1) print(scores2)
```

Listing 5.20. Three-layer LSTM for prediction of the next price movement direction.

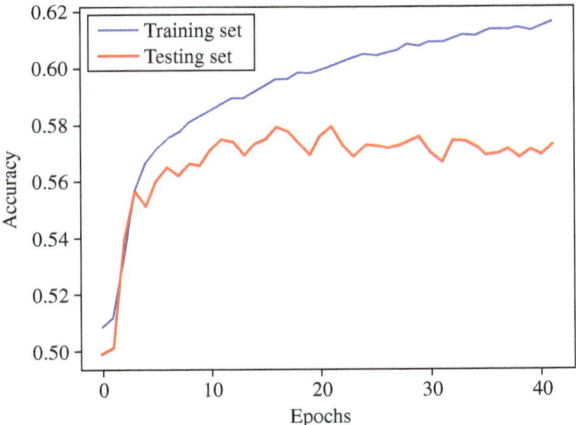

Figure 5.45. The accuracy of a three-layer LSTM w.r.t. the number of epochs in the training and testing sets.

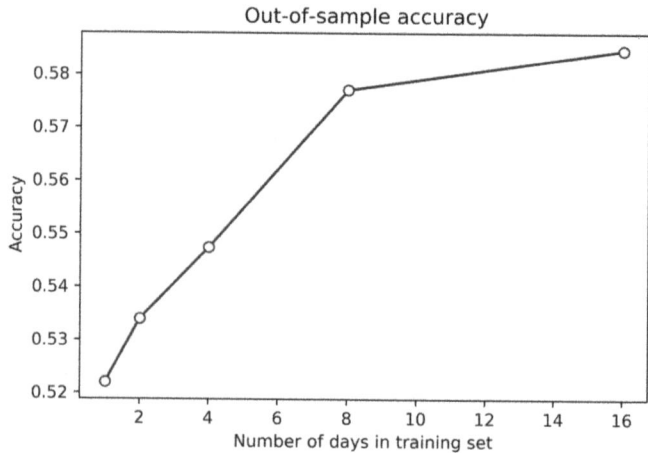

Figure 5.46. Plot of out-of-sample accuracy vs. number of days in the training set.

5.5 Exercises

(1) Write down the mathematical formulation of the neural network model in Figure 5.7.
(2) Apply LSTM to the IMDB dataset, which is a standard movie review dataset for sentiment analysis[23] (You may need to use word embedding to transform the text data into a numeric vector).

---

[23]You can download the IMDB data following the instructions in `https://keras.io/datasets/#imdb-movie-reviews-sentiment-classification`.

Chapter 6

Cluster Analysis

In the supervised learning chapters, we have seen machine learning algorithms based on the assumptions or model of data. In this chapter, we focus on cluster analysis as another type of algorithm aimed at exploring the general and natural characteristics of multidimensional data without pre-specified data model assumptions.

6.1 Introduction

Cluster analysis, also known as clustering, is to discover structure in data in terms of natural grouping based on the homogeneity of the data. In other words, we group samples into clusters such that samples in the same cluster are more similar than samples in different clusters.

It looks similar to the classification problem in supervised learning as both involve categorizing the data. However, unlike the classification problem, which tries to use observations of variables X to predict the categorical response variable Y, cluster analysis aims to discover the structure of X itself without the response variable Y.

Cluster analysis has been applied in many areas for a long time. It has become especially important today since more high volume and high dimensional datasets have become available through the advance of technology. Clustering can help understand and process data due to its exploratory nature. Numerous applications include image segmentation in computer vision, efficient information processing from documents, studying behavior, and characterizing people for social science study and commercial use. Clustering has also been applied in finance, and we will see some examples later in this chapter.

In Section 6.2, we first introduce the general clustering framework and basic concepts. From Sections 6.3 to 6.6, we discuss different types of clustering methods, including partitional clustering (K-means), hierarchical clustering, density-based clustering (DBSCAN) and distribution-based clustering. Python implementation examples can be found in Section 6.7, and applications in finance are in Section 6.8.

6.2   Clustering Framework

Before introducing different clustering algorithms, we first explain the general clustering framework by answering the following questions:

- What is the objective of clustering problems?
- How do we describe the similarity between two data objects?
- What are the available clustering methods?
- How do we evaluate clustering results?

6.2.1   Data and objective

We use the same notations defined in supervised learning. Recall that the data X in Section 2.1 are a set of data samples, represented by N observations measured on p features $X_1, \ldots, X_p$. For clustering, we group the data X into a set of clusters $C = \{C_1, \ldots, C_k\}$, such that data objects in the same cluster are more similar to each other than the data objects in other clusters.

The set of clusters C contains all data objects. The assignment of data object i to cluster C_k is a mapping from the object to a cluster c: $c(i) = C_k$. Each data object can belong to one or more clusters, and when it belongs to a cluster, the belonging can be expressed with certainty or with likelihood. Clustering with certainty is called hard-clustering, and clustering with probability is called soft-clustering, which we will see in the distribution-based clustering section. In this chapter, we only discuss the case where each data object belongs to one and only one cluster.

Since the data consists of observations of features, then depending on in which dimension we aim to find more structural information, we can cluster either observations or features. For example, in Figure 6.1, when we look into the stock market, we collect a set of historical stock data featured by the market capital, price, return, trading volume, volatility and some other features. In order to know whether some stocks are more similar than others, we can cluster stocks into subgroups based on the features.

clustering observations based on features

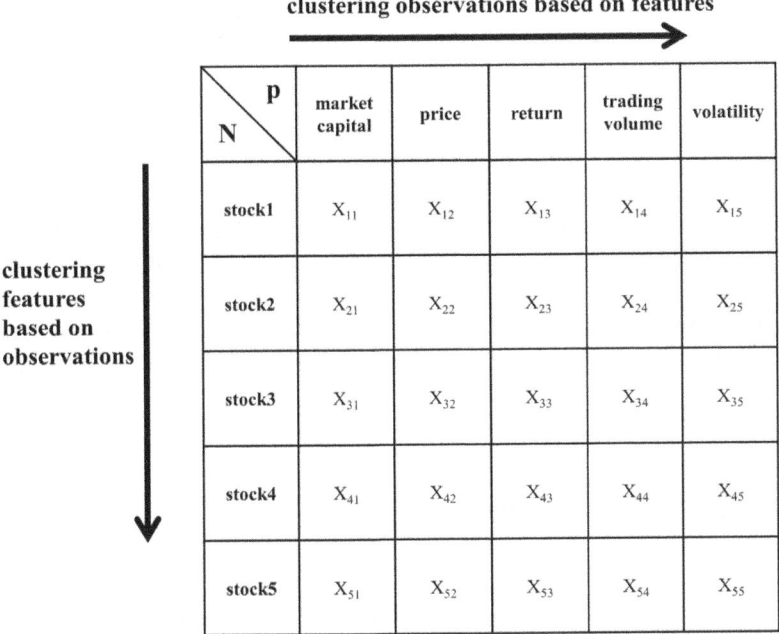

N \ p	market capital	price	return	trading volume	volatility
stock1	X_{11}	X_{12}	X_{13}	X_{14}	X_{15}
stock2	X_{21}	X_{22}	X_{23}	X_{24}	X_{25}
stock3	X_{31}	X_{32}	X_{33}	X_{34}	X_{35}
stock4	X_{41}	X_{42}	X_{43}	X_{44}	X_{45}
stock5	X_{51}	X_{52}	X_{53}	X_{54}	X_{55}

clustering features based on observations

Figure 6.1. The choice of clustering direction.

On the other hand, if we need to explore feature representation revealing the structural information, we can cluster features based on the observations. If not mentioned explicitly, in this chapter, we will focus on clustering observations based on features.

6.2.2 Similarity measures

From the objective above we can see that the clustering process begins with the question: how similar are two objects? The objects can be either individual data points or data groups. In the following, we introduce the similarity between data points, and group similarity can be defined based on this, which we will see in Section 6.4.1.

Similarity: for any $x, y \in X$, the similarity measure is a distance function $d : X \times X \to [0, \infty)$ satisfying:

(1) non-negativity $d(x, y) \geq 0$;
(2) symmetry $d(x, y) = d(y, x)$;
(3) triangle inequality $d(x, z) \leq d(x, y) + d(y, z)$;
(4) $d(x, y) = 0 \Rightarrow x = y$.

Suppose we cluster N observations based on p features. For samples i and i', denote the similarity on feature j as $d_j(x_{ij}, x_{i'j})$; then we define the similarity based on all features by

$$d(x_i, x_{i'}) = \sum_{j=1}^{p} w_j d_j(x_{ij}, x_{i'j}),$$

where $w_j \geq 0$ is the weight assigned to feature j representing the relative importance of the feature j. The weights assignment depends on the context of the data representation and the goal of data segmentation. For simplicity, we adopt equal weights here, and the distance measures are taken equally among all features, so that the distance function d can be written as:

$$d_{ii'} := d(x_i, x_{i'}) = \sum_{j=1}^{p} d(x_{ij}, x_{i'j}).$$

The following distance functions are based on different norms and are widely used in clustering:

- Euclidean distance: $d(x_i, x_{i'}) = ||x_i - x_{i'}||_2 = \sum_{j=1}^{p} |x_{ij} - x_{i'j}|^2$
- Manhattan distance: $d(x_i, x_{i'}) = ||x_i - x_{i'}||_1 = \sum_{j=1}^{p} |x_{ij} - x_{i'j}|$
- Chebyshev distance: $d(x_i, x_{i'}) = ||x_i - x_{i'}||_\infty = \max_{j=1,\ldots,p} |x_{ij} - x_{i'j}|$

6.2.3 Clustering methods

Partitional clustering

Partitional clustering assigns each data object to a cluster based on similarity. The framework is to iteratively generate clusters by optimizing the objective functions until convergence. The most common method is K-means, introduced in Section 6.3. Like K-means, most of the partitional clustering methods need a pre-specified number of clusters. The variations can be from different similarity measures, initialization and so on.

Hierarchical clustering

Hierarchical clustering recursively builds nested clusters in either a top-down or bottom-up approach, which reveals more information than partitional clustering. Hierarchical clustering has a tree representation, while we only see the final clustering results with a flat structure in partitional clustering methods. We will introduce the bottom-up structure in Section 6.4. The variation is mainly from different assumptions about the linkage, which is the group similarity measure.

Density-based clustering

Density-based clustering groups data objects into clusters and outliers based on the concept of density.

Distribution-based clustering

Distribution-based clustering is also called soft-clustering. Unlike the previous deterministic approaches, it assigns data objects to clusters with probability.

6.2.4 Clustering validation

After obtaining clustering results, it is crucial to assess the performance as results can be informative if the formed clusters have clear structures and patterns. There are two types of metrics to provide solutions:

- If there is no external information about the natural grouping, we can measure how well clusters are separated and how compact the formed clusters are. *Silhouette width* is a standard metric in this category, which we will introduce below.
- If we have external reference labels for the data, we can assess the clustering quality by comparing the cluster labels with the reference ones. The goal is not to count different labeling from the reference as an error but to assess whether clustering results are similar to the reference labels. Common metrics include the adjusted Rand index (ARI).

Silhouette width

The silhouette width can help us to distinguish weak and clear-cut clusters. Define the silhouette width $s(i)$ of data object i as

$$s(i) := \frac{b(i) - a(i)}{max(a(i), b(i))}, \tag{6.1}$$

where

- $a(i)$ is the average distance between data item i and all other data within the same cluster.
- $b(i)$ is the average distance between data item i and the nearest neighbor cluster.

From the definition, we can see $s(i)$ takes values in $[-1, 1]$, and

- If $s(i)$ is close to 1, this means that the within-cluster distance $a(i)$ is small and the distance to neighbour $b(i)$ is big. Therefore the data is appropriately clustered.
- If $s(i)$ is close to zero, this means that $a(i)$ and $b(i)$ are close to each other, and the data item i is on the border of two clusters.

- If $s(i)$ is close to -1, this is due to the within-cluster distance for data item i being much larger than the distance from i to the neighbour, so this implies a poor assignment for data item i.

With the metric defined for each data point, we can evaluate the overall clustering quality by taking the average of the metric of data points from each cluster over the whole dataset. In Section 6.3, we can see that the silhouette width provides a graphical representation of clustering quality. In addition, it can be used to choose the optimal number of clusters.

6.3   *K*-means

6.3.1   Introduction

In this section, we introduce the most commonly used partitional clustering method: K-means. The goal is to find K non-overlapping clusters from the given data set. In K-means, the similarity measure is normally taken as the Euclidean distance, and the objective is to find the clusters $C = \{C_1, \ldots, C_K\}$ by solving

$$\underset{\{C_1,\ldots,C_k\}}{\mathrm{argmin}} \sum_{k=1}^{K} \sum_{\{x_i \in C_k\}} ||x_i - \mu_k||^2,$$

where μ_k is the centroid of the cluster C_k: $\mu_k = \frac{1}{|C_k|} \sum_{x_i \in C_k} x_i$, and $|C_k|$ is the size of the cluster C_k.

As the total variance $\sum_{x_i, x_{i'} \in C} ||x_i - x_{i'}||^2$ is fixed, and the total variance is the sum of the within-cluster variance and the between-cluster variance, therefore the above objective function is equivalent to minimizing the within-cluster variance:

$$\underset{\{C_1,\ldots,C_k\}}{\mathrm{argmin}} \sum_{k=1}^{K} \frac{1}{2|C_k|} \sum_{\{x_i, x_{i'} \in C_k\}} ||x_i - x_{i'}||^2.$$

The solution is found via an iterative process described in Algorithm 10.

Algorithm 10: K-means Clustering

1: Select an initial set of K cluster centers.
2: Generate clusters by assigning each data to its closest cluster center.
3: Update cluster centers.
4: Iterate steps 2 and 3 until no change in the assignment.

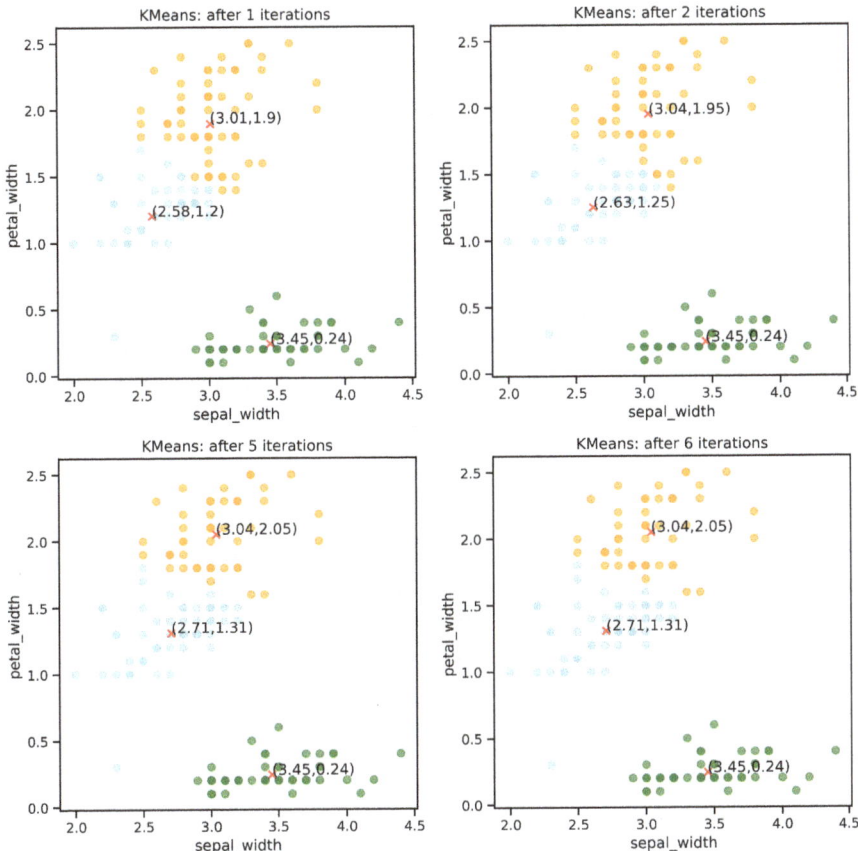

Figure 6.2. *K*-means iterations on the Iris dataset.

Here we demonstrate with samples of Iris data where the selected features are petal width and sepal width. The objective is to cluster the observations into three clusters via *K*-means. The iteration to find the solution is shown in Figure 6.2, and clusters are in different colors. We can see that the solution has been found after five iterations. Note that the algorithm converges to a local optimal solution instead of a global one.

6.3.2 Practical issues

In the *K*-means algorithm, (1) initialization of clusters, (2) distance metric and (3) number of clusters *K* are the user inputs, and we look at how to chose those inputs in practice.

6.3.2.1 *Initialization of clusters*

The K-means algorithm above starts with random initialization of cluster centers. Different initialization values result in different clustering results. In practice, we can do clustering multiple times with different initialization and take the result from the best evaluation result.

6.3.2.2 *Distance metric*

Euclidean distance indicates that the clusters from K-means are spherical, so the clusters from K-means tend to be ball-shaped. Therefore K-means is unable to find clusters of other shapes. Extensions of K-means include other types of distance such as Mahalanobis distance and Manhattan distance.

6.3.2.3 *Number of clusters K*

The number of clusters K is a pre-defined parameter in the K-means algorithm. As we can see in Figure 6.3, different K values yield different clustering results. Choosing a too large or too small K can lead to undesired clustering results. How to choose the parameter is an important question. Either we can find some guidance from the context of the application, or we apply the selection criteria that we introduce in the following.

- Elbow method

  The method can be traced back to a paper by R. Thorndike [Thorndike (1953)]. Recall that the goal of K-means is to minimize the within-cluster variance (WCV), and WCV is a decreasing function of the number of clusters. On the other hand, having too many clusters may overlook the homogeneity of the data. Therefore, the elbow method tries to find K at which adding another cluster does not reduce WCV by much. If we plot the WCV against the number of clusters K, then K at the bend in the plot where the decrease in WCV becomes much smaller when K increases is considered to be an appropriate number of clusters. For example, in Figure 6.4, the elbow is around $K = 6$.

- Average silhouette width

  As mentioned in Section 6.2, the silhouette width $s(i)$ can be used to evaluate the clustering results. Here it also can be used to determine the number of clusters K.

  If K is set to too low for data with dense clusters far away from each other, then the clustering algorithm will tend to combine some clusters

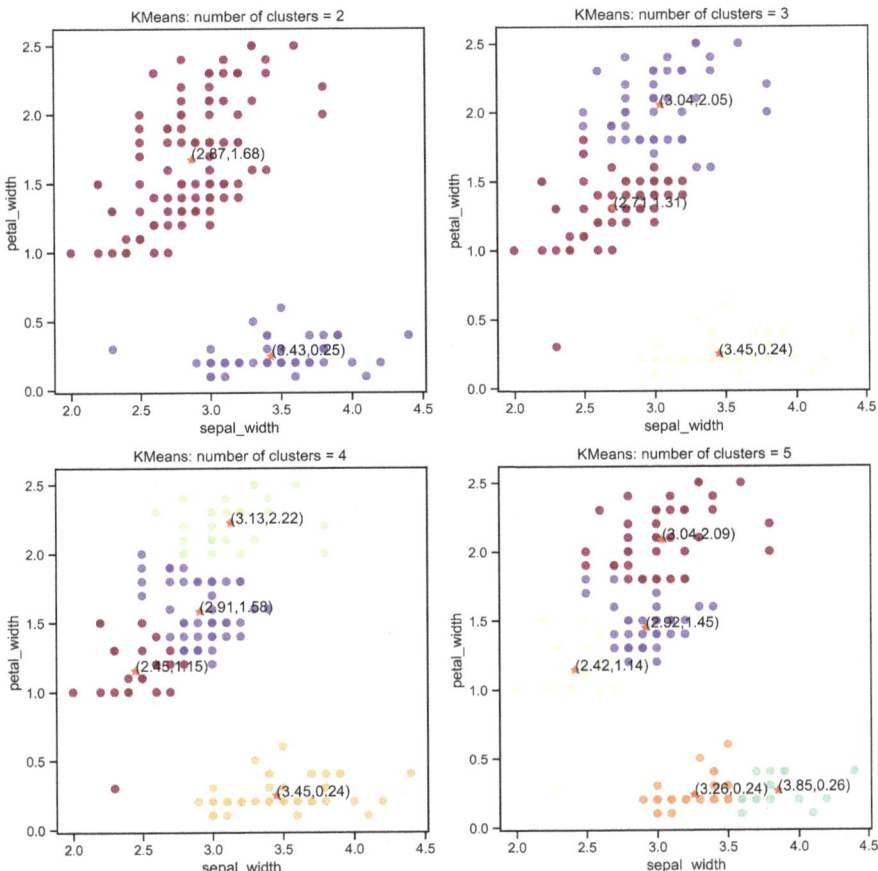

Figure 6.3. *K*-means: choosing a different number of clusters *K*.

to reduce the number of clusters. In this case, we will have a large within cluster dissimilarity/distance $a(i)$ and a small $s(i)$ value.

If K is set to too high, then we have the natural clusters divided into subgroups. Then, the distance between neighboring clusters $b(i)$ tends to be small, which results in a small $s(i)$ value.

From the above, we can see that the optimal K should make the silhouettes as wide as possible. Figure 6.5 gives an example of this method.

• Other methods

As mentioned at the beginning, one can set different criteria to determine K—for example, the gap-statistics proposed by R. Tibshirani *et al.* [Tibshirani *et al.* (2001)]. In addition, sometimes visualization

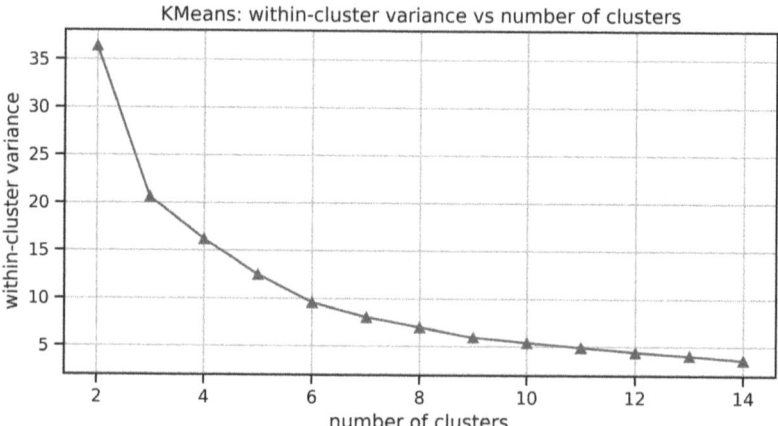

Figure 6.4. K-means: within-cluster variance as a function of the number of clusters.

can be very helpful, and it can be used either before or after other methods for clustering validation.

6.3.3   Summary

- The K-means clustering method aims to find clusters minimizing the within-cluster variance.
- The iterative K-means algorithm terminates at a local optimal solution, and the clustering results can strongly depend on the initial assignment.
- The Euclidean distance leads to ball-shaped clusters so K-means cannot be used to discover particularly non-convex shaped clusters.
- K-means needs to specify the number of clusters K in advance though one can use some criteria to decide K based on either domain knowledge or clustering results.

6.4   Hierarchical Clustering

In contrast to the K-means method, in this section we introduce hierarchical clustering, which provides the clustering results in a tree structure.

There are two types of hierarchical clustering. One is *hierarchical agglomerative clustering* (HAC), which is a bottom-up approach starting from all singleton clusters[1] and merges pairs of clusters recursively until all

---

[1] Each data object is a cluster.

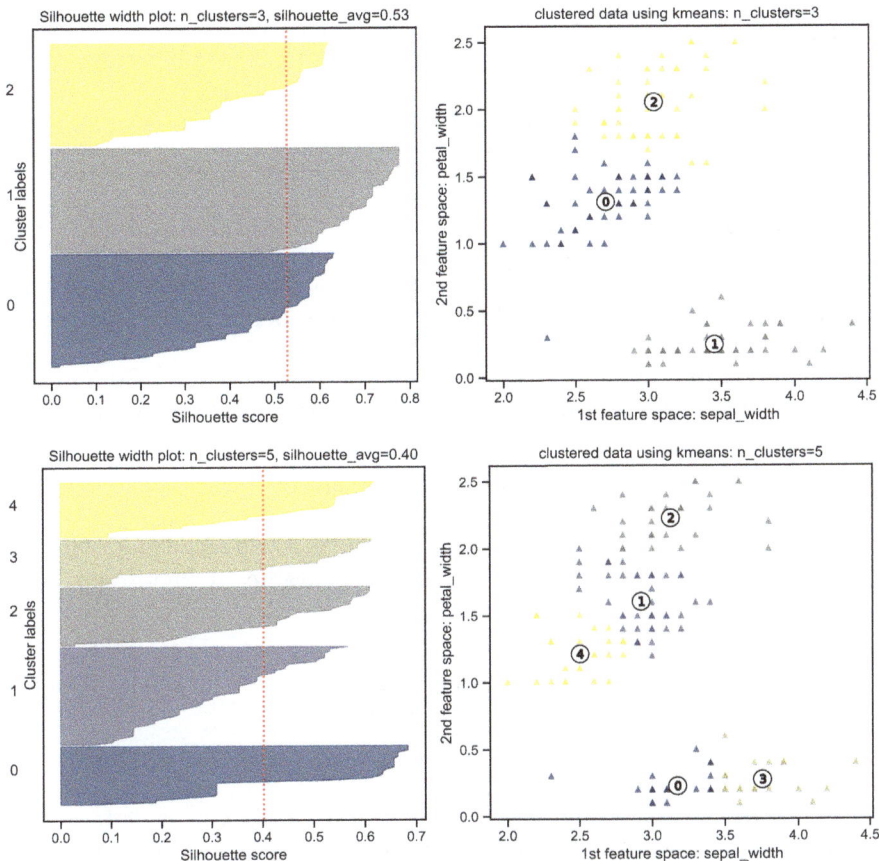

Figure 6.5. An example of choosing the optimal K from average silhouette width.

data objects are in the same cluster. The other type is *hierarchical divisive clustering* (HDC); this takes a top-down approach that starts with a cluster including all data objects and splits clusters into pairs recursively until every data object becomes a singleton cluster.

Note that the structure above is naturally linked to a tree representation where HAC starts from all leaves and combines leaf clusters down to the root, while HDC starts from the root and splits into branches until it has clusters with each consisting of a single leaf. In this section, we focus on the bottom-up approach (HAC) and see how the HAC performs the clustering using a dendrogram (diagram of a tree structure). The iterative process of HAC mentioned above is shown in Algorithm 11.

Algorithm 11: Hierarchical Agglomerative Clustering (HAC)

1: Start with all data objects and treat each data object as a cluster.
2: Compute the pairwise group similarity for all pairs of clusters.
3: Merge two of them having the minimal group distance.
4: Iterate steps 2 and 3 until all data objects are in the same cluster.

6.4.1 Linkage

Note that in the HAC algorithm above, we need to choose the *linkage* (group similarity). In the following, we introduce three types of group linkages.

- Single Linkage (SL)
  SL takes the minimal distance between any two points in group G and H, i.e.,

$$d_{SL}(G, H) = \min_{x_i \in G, x_j \in H} d(x_i, x_j).$$

  As SL is determined by the similarity between two single data objects irrespective of other objects, it does not guarantee within-group similarity.

- Complete Linkage (CL)
  Instead of the minimal distance, CL takes the maximal distance between two data objects:

$$d_{CL}(G, H) = \max_{x_i \in G, x_j \in H} d(x_i, x_j).$$

  Using a different group distance measure from SL, CL tends to merge clusters with similar diameters and produce compact clusters. Because of the maximal distance, it could happen that data objects to be merged are closer to members of other clusters than to their within-cluster members.

- Group Average (GA)
  This uses the average dissimilarity over all pairs of data objects in these two groups:

$$d_{GA}(G, H) = \frac{1}{N} \sum_{x_i \in G, x_j \in H} d(x_i, x_j) = \frac{1}{|G|} \frac{1}{|H|} \sum_{x_i \in G, x_j \in H} d(x_i, x_j),$$

  where $|G|$ and $|H|$ are the size of the groups. By taking the average, GA can be seen as a compromise between SL and CL.

Besides the above linkages, there are also other alternatives such as the centroid-based linkage, which is based on the distance of cluster centers pairs: $d_C(G, H) = d(\mu_G, \mu_H)$.

In addition, we mention here Ward's method, which relies on the within-cluster variance change when merging. Ward's method was proposed by Ward, J.H. Jr. [Ward Jr (1963)]. It recursively merges two clusters by selecting the cluster to merge such that the increase in within-cluster variance is minimal. The distance metric used is the Euclidean distance.

6.4.2 Dendrogram

The dendrogram is a tree-type diagram that can be used to visualize the hierarchy of clusters in HAC. For example, Figure 6.6 shows how clusters are formed from singleton clusters to the one big cluster, including all data objects using different linkages for data.

While the dendrogram itself provides us with the dynamic of cluster forming, one needs to decide at which level in the tree to stop as the final clustering result. One can specify the variance threshold in the dendrogram, which is equivalent to specifying the number of clusters. Figure 6.7 shows that after specifying the variance threshold in each linkage case, the number of clusters is three. We can see that the clusters are very different.

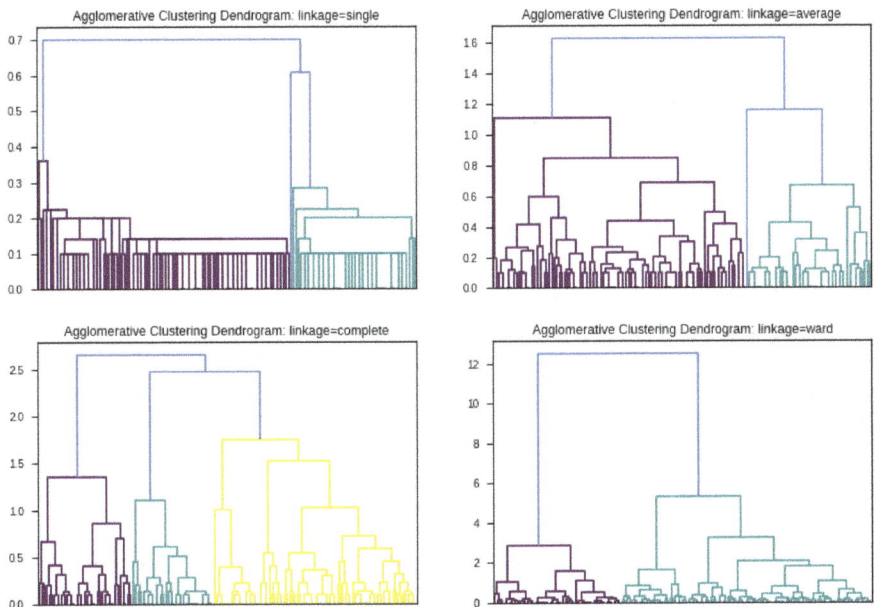

Figure 6.6. Dendrogram for hierarchical agglomerative clustering with different linkages. The vertical axis shows the Euclidean distance between clusters.

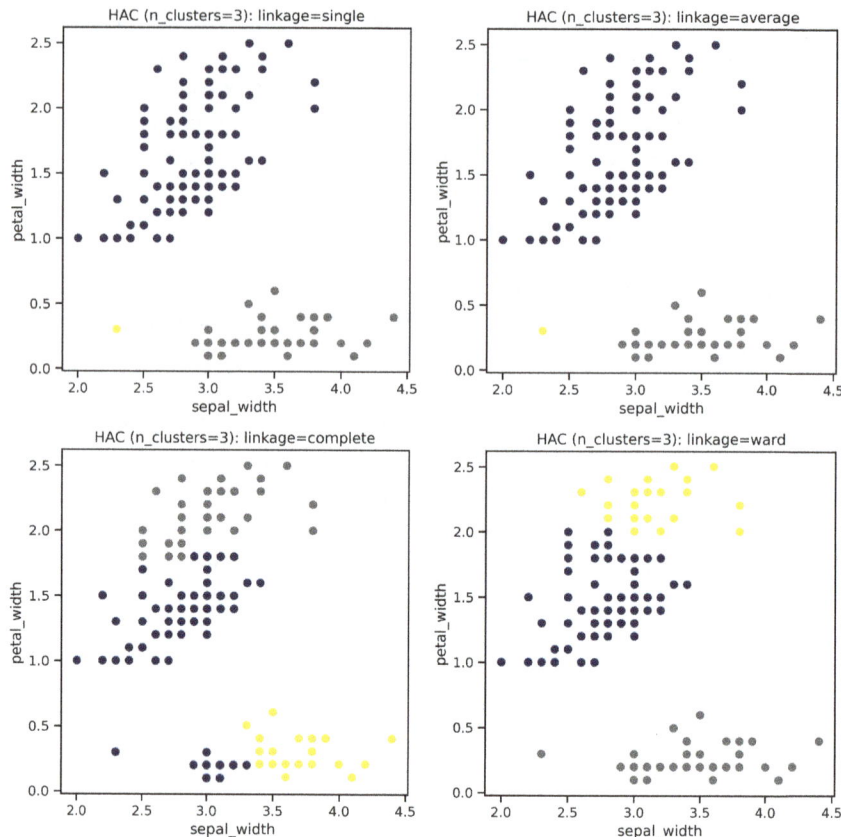

Figure 6.7. HAC with a certain variance threshold.

6.5 Density-based Clustering: DBSCAN

The density-based clustering methods try to find different density regions and group data objects in those regions. In this section, we introduce the DBSCAN method as an example.

6.5.1 Definition

DBSCAN (Density-Based Spatial Clustering of Applications with Noise) is a clustering algorithm proposed by Ester *et al.* [Ester *et al.* (1996)]. It is a density algorithm as the concept of clusters is based on the area density and the data points' connectivity. See Figure 6.8 as an example using DBSCAN and compared with Ward's method that we have seen in

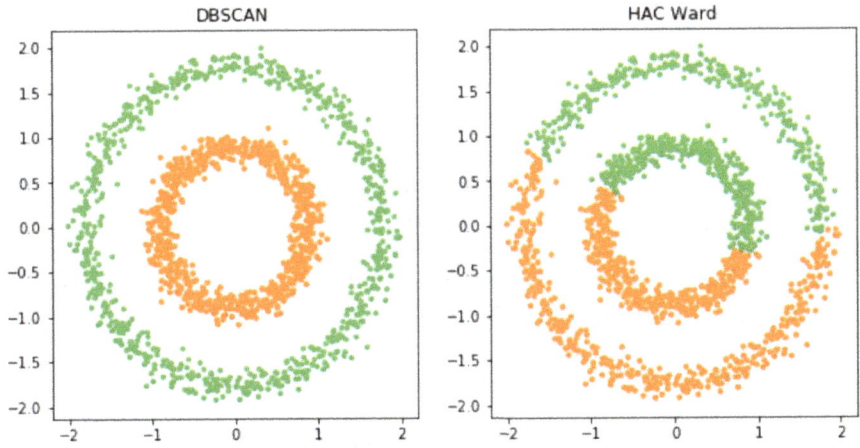

Figure 6.8. DBSCAN example.

hierarchical agglomerative clustering. In the following, we first introduce some basic concepts.

The concept *density* is characterized by a parameter pair $\theta = (\epsilon, minPoints)$ and a distance metric d, where ϵ is a positive real number describing a radius and $minPoints$ is a positive integer. The parameter θ describes the density concept around a point, and the core point condition characterizes the density criterion:

- ϵ-**neighborhood** of point $p \in X$: $N_\epsilon(p) := \{q \in X | d(p,q) \leq \epsilon\}$.
- **core point condition** w.r.t. θ: a point $q \in X$ is a core point if $N_\epsilon(q) \geq minPoints$.

Let us take a look at the concept of *connectivity*, based on which the clustering will be introduced. For two points p and q in X:

- p is **directly density-reachable** from q w.r.t. θ: if $p \in N_\epsilon(q)$ where q is a core point of X w.r.t. θ.
- p is **density-reachable** from q w.r.t. θ: if there exists a chain of points $p_1 = p, p_2, \ldots, p_n = q$ in X, such that for all $i = 1, 2, \ldots, n-1$, p_{i+1} is directly reachable from p_i.
- p and q are **density-connected**: if there exists a point o in X such that both p and q are density-reachable from o.

We can see that density-reachable is a weaker condition than directly density-reachable. Moreover, unlike the density-reachable condition, the density-connected condition is symmetric.

A **cluster** C in X w.r.t. $\theta = (\epsilon, minPoints)$ is a non-empty subset including core points and all points that are density-reachable from those core points. Once all clusters $\{C_i\}_i$ are found for X, the **noise** of X is the set of points that are not in any cluster.

Note that each cluster can have its own parameter θ, but in practice, a global one is often specified for all clusters.

The cluster defined above satisfies the following properties:

(1) For any $p, q \in C$, p is density-reachable from q w.r.t. θ. In other words, if a point is density-reachable from any point in the cluster, then it belongs to the same cluster.
(2) All points in the same cluster are mutually density-connected.

The algorithm is summarized as follows in Algorithm 12.

Algorithm 12: DBSCAN Algorithm

Given $\theta = (\epsilon, minPoints)$:

  1: Find all core points of the dataset.
  2: For each core point, find a cluster by retrieving all points that are density-reachable from this core point.
  3: After 1 and 2, the remaining unassigned points in the dataset are assigned to the noise set.

From the definition, we can see that a cluster contains two types of points. The first type is the core point satisfying the core points condition. The second type we call a border/edge point, which usually contains fewer points in its N_ϵ neighborhood than the core points.

6.5.2   Determine parameters

The challenge using DBSCAN is to choose appropriate parameters: $\theta = (\epsilon, minPoints)$ and distance d.

Density parameter $minPoints$**:** The original paper [Ester *et al.* (1996)] suggests setting $minPoints = 2 \times p - 1$ where p is the feature dimension of the dataset.

Density parameter ϵ**:** Too small ϵ results in a large part of the data not being clustered, and too large ϵ leads to the majority of the data being in the same cluster. One can use a k-distance graph to give hints of the density distribution: choosing $k = minPoints - 1$, plotting in descending order of the distance to the k-th neighborhood for all points.

Select ϵ on the y-axis at the "elbow" in the graph as intuitively this indicates the boundary between noise and points in clusters.

Distance metric d: One should choose the distance metric before deciding the other parameters. The choice depends on the dataset and application context in general, and Euclidean distance is the most common choice.

6.5.3 Advantage of DBSCAN

Based on the connectivity concept, the DBSCAN algorithm and the observation of Figure 6.8, we can summarize the advantage of using DBSCAN as follows:

- DBSCAN distinguishes clusters and noise based on the connectivity concept, therefore compared to other clustering methods introduced earlier, it is more robust to outliers.
- Unlike other clustering methods that depend on the initialization of the algorithm, DBSCAN is almost a deterministic clustering, though the border points of DBSCAN clusters may depend on the order of data if two clusters are very close to each other.
- DBSCAN can find clusters with challenging shapes of data compared with other clustering methods like K-means and HAC. Figure 6.8 is an example.
- DBSCAN does not require specifying the number of clusters a priori like K-means.

6.6 Distribution-based Clustering

6.6.1 Introduction

In previous sections, different clustering methods assigned each object to a cluster or a noise set. As the assignment have no uncertainty, we call those methods **hard-clustering**. Unlike hard-clustering, **soft-clustering** assigns data objects to clusters with a probability. Therefore it is also called a distribution-based clustering method. In other words, one data object can belong to multiple clusters.

The soft-clustering result has the following representation:

$$\mathbb{P}(x_i \in C_k) = p_{k,i} \text{ for any } x_i \in X, \text{ and } \sum_k p_{k,i} = 1,$$

i.e., the probability for x_i to be assigned to the final cluster set $C = \{C_k\}_k$ is 1.

Soft clustering assumes a probability distribution from which the dataset is generated. A typical distribution assumption is the **Gaussian mixture model (GMM)**. A K-*Gaussian mixture distribution* is a linear combination of K Gaussian distributions

$$p(x) = \sum_{k=1}^{K} \pi_k p_N(x|\mu_k, \Sigma_k), \tag{6.2}$$

where the mixing coefficient $\pi_k \in [0,1]$ for any k in $\{1, \ldots, K\}$, and $\sum_{k=1}^{K} \pi_k = 1$.

Denote the density of the k-th Gaussian component X_k as $p_N(x|\mu_k, \Sigma_k)$, and the Gaussian mixture density can be written as

$$p(x) = \sum_{k=1}^{K} \pi_k p_N(x|\mu_k, \Sigma_k). \tag{6.3}$$

Note that by choosing a sufficiently large number of components K and adjusting the parameters π, μ, Σ, almost all continuous density functions can be approximated by the Gaussian mixture family with a given accuracy level. Therefore the GMM is widely used in applications. Since the Gaussian mixture density is determined by the parameter set $\pi = (\pi_k)_k$, $\mu = (\mu_k)_k$ and $\Sigma = (\Sigma_k)_k$, the challenge is to estimate the parameters by maximizing the density function.

If we use the maximum likelihood method for estimation, then we can write the log likelihood function as

$$\ln p(X|\pi, \mu, \Sigma) = \sum_{n=1}^{N} \ln \left(\sum_{k=1}^{K} \pi_k p_N(x_n|\mu_k, \Sigma_k) \right). \tag{6.4}$$

Note that the problem above does not have a closed-form solution as the summation $\sum_{k=1}^{K} \pi_k p_N(x_n|\mu_k, \Sigma_k)$ is inside the logarithm. One powerful tool we can use is the **expectation–maximization** (EM) algorithm.

We first introduce the general framework of EM in Section 6.6.2 and then focus on the GMM as a special case in Section 6.6.3.

6.6.2   Expectation–maximization algorithm

Denote the parameter set as θ; from Section 6.6.1 we know the goal is to solve $\theta^* = \mathrm{argmax}_\theta \, p(X|\theta)$ where $p(X|\theta)$ is defined in (6.4). The EM algorithm transforms this maximization problem into a different one by introducing some latent (unobserved) variables Z. The complete data after augmentation is (X, Z) instead of the original data X.

Given the parameter θ, the joint distribution of (X, Z) is $p(X, Z|\theta)$. The key to the EM algorithm is: (1) we assume that maximizing the likelihood of the complete data (X, Z) is easier than for the original likelihood of X only; (2) the solution that maximizes $p(X, Z|\theta)$ is also the solution to the original maximization problem.

From Bayes' theorem we have:

$$p(X|\theta) = \frac{p(X, Z|\theta)}{p(Z|X, \theta)}$$

$$p(X|\theta) = \sum_Z p(X, Z|\theta).$$

Assume an arbitrary distribution $q(Z)$ for the latent variable Z; from the equations above, the log-likelihood function can be written as

$$\ln p(X|\theta) = \underbrace{\sum_Z q(Z) \ln \frac{p(X, Z|\theta)}{q(Z)}}_{:=L(q,\theta)} + \underbrace{\left(-\sum_Z q(Z) \ln \frac{p(Z|X, \theta)}{q(Z)} \right)}_{:=KL(q,p,\theta)}, \quad (6.5)$$

where $KL(q, p, \theta)$ is the *Kullback–Leibler divergence* between distributions $q(Z)$ and $p(Z|X, \theta)$ satisfying $KL(q, p, \theta) \geq 0$.

From Equation (6.5), it is clear that $\ln p(X|\theta) \geq L(q, \theta)$, and $\ln p(X|\theta) = L(q, \theta)$ if and only if the posterior distribution equals the distribution of Z, i.e., $p(Z|X, \theta) = q(Z)$.

Algorithm 13: EM Algorithm

1: Initialization: Choose an initial parameter set θ_{old}.
2: Expectation Step (*E step*): fix θ_{old}; maximize $L(q, \theta_{old})$ over $q(Z)$:

$$q^*(Z) = \underset{q}{\mathrm{argmax}}\, L(q, \theta_{old}). \quad (6.6)$$

The solution is $q^*(Z) = p(Z|X, \theta_{old})$ from Equation (6.5).
3: Maximization Step (*M step*): using $q^*(Z)$ obtained from the E step, maximize $L(q, \theta)$ with respect to θ:

$$\theta_{new} = \underset{\theta}{\mathrm{argmax}}\, L(q^*, \theta) = \underset{\theta}{\mathrm{argmax}} \sum_Z p(Z|X, \theta_{old}) \ln p(X, Z|\theta). \quad (6.7)$$

4: Convergence Check: if the solution θ_{new} does not satisfy the pre-defined convergence criterion, then set θ_{old} with the value of θ_{new} and repeat E step and M step in 2 and 3.

Based on the above decomposition, we introduce the EM Algorithm, which is an iterative process, and it maximizes $\ln p(X|\theta)$ in Equation (6.5). The algorithm details can be found in Algorithm 13.

- After each iteration in (E step, M step), the solution θ_{new} from the M step increases the logarithm function $\ln p(X|\theta)$, since
  - from the E step, we have $\ln p(X|\theta_{old}) = L(q^*, \theta_{old})$;
  - from the M step, $\ln p(X|\theta_{new}) = L(q^*, \theta_{new}) + KL(q^*, p, \theta_{new}) \geq L(q^*, \theta_{new})$.

  Since $L(q^*, \theta_{new}) \geq L(q^*, \theta_{old})$, we have $\ln p(X|\theta_{new}) \geq \ln p(X|\theta_{old})$. The convergence follows from the monotonicity.
- The main optimization problem is at the M step. The maximization problem in Equation (6.7) is the summation of logarithm functions, which is in general an easier problem than the original problem in Equation (6.4).
- Note that Equation (6.7) holds as the entropy $-q^*(Z) \ln q^*(Z)$ is a function of θ_{old} only, thus it was removed from the objective function.
- The EM algorithm does not guarantee to converge to a global maximum, and it typically converges to a local maximum.

6.6.3   Gaussian mixture model

Here, the latent variables Z are the mixing variables in the GMM as mentioned in Equation (6.2). We summarize the algorithm in Algorithm 14. An example of clustering using GMM can be found in Figure 6.9.

6.7   Clustering with Python

The scikit-learn library contains the *sklearn.cluster* module, which performs clustering on unlabeled data. We show briefly here some examples of how to call clustering methods; the complete code can be found in the Github repository accompanying to the book.

6.8   Numerical Example

The implementation for K-means, Dendrogram and HAC can be found in Listings 6.21, 6.22 and 6.23, respectively.

Algorithm 14: EM for Gaussian Mixture

For each $k = 1, 2, \ldots, K$:

1: Initialization: take initial values for parameters $(\mu_k^{old}, \Sigma_k^{old}, \pi_k^{old})$.

2: Expectation Step (E step): evaluate the posterior probability of Z given X also called the responsibilities r:

$$
\begin{aligned}
r(z_{nk}) &= p(Z_k = 1 | X_n) \\
&= \frac{p(Z_k = 1)p(X_n | Z_k = 1)}{\sum_{j=1}^{K} p(Z_j = 1)p(X_n | Z_j = 1)} \\
&= \frac{\pi_k^{old} p_N(X_n | \mu_k^{old}, \Sigma_k^{old})}{\sum_{j=1}^{K} \pi_j^{old} p_N(X_n | \mu_j^{old}, \Sigma_j^{old})}.
\end{aligned}
\tag{6.8}
$$

3: Maximization Step (M step): with the responsibilities above, denote $N_k := \sum_{n=1}^{N} r(z_{nk})$, and update the parameters below that maximize the objective function $L(r, \theta)$:

$$
\mu_k^{new} = \sum_{n=1}^{N} \frac{r(z_{nk})}{N_k} X_n,
$$

$$
\Sigma_k^{new} = \sum_{n=1}^{N} \frac{r(z_{nk})}{N_k} (X_n - \mu_k^{new})(X_n - \mu_k^{new})^T,
$$

$$
\pi_k^{new} = \frac{N_k}{N}.
$$

4: Convergence Check: with the updated parameters $(\mu^{new}, \Sigma^{new}, \pi^{new})$ from the M step, check whether the log-likelihood $\ln p(X | \mu^{new}, \Sigma^{new}, \pi^{new})$ converges or not. If not, go back to step 2.

```python
from sklearn.cluster import KMeans

def model_kmeans(data,n_clusters,init_method='k-means++'):
    model = KMeans(n_clusters=n_clusters,
                   init=init_method)
    model.fit(data)
    cluster_labels = model.predict(data)
    cluster_centers = model.cluster_centers_
    within_cluster_variance = model.model.inertia_
    fit_res = {'cluster_method':'kmeans',
               'model':model,
               'cluster_labels':cluster_labels,
               'cluster_centers':cluster_centers,
               'cluster_wcv':within_cluster_variance}
    return fit_res
```

Listing 6.21. *K*-means.

Gaussian Mixture

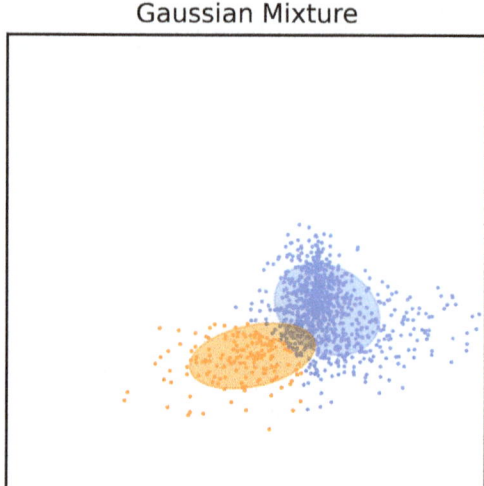

Figure 6.9. Gaussian mixture model with two components and predicted clustering labels.

```
1  from scipy.cluster import hierarchy
2  # linkage input options: 'single','average','complete','ward'
3  linkage = 'complete'
4  fig = hierarchy.dendrogram(Z=hierarchy.linkage(data, linkage))
```

Listing 6.22. Dendrogram for hierachical agglomerative clustering.

```
1  from sklearn.cluster import AgglomerativeClustering
2
3  def agglomerative_clustering(data,linkage,n_clusters):
4    model = AgglomerativeClustering(linkage=linkage,
5                                    n_clusters=n_clusters)
6    model.fit(data)
7    labels = model.labels_
8    fit_res = {'labels':labels,'fitted_model':model}
9    return fit_res
```

Listing 6.23. HAC with given number of clusters.

Applications can be found in Section 7.4.5, where we apply clustering to PCA.

6.9 Exercises

(1) What is the main difference between clustering and classification?
(2) Enumerate three methods of clustering and explain how they work.

Chapter 7

Principal Component Analysis

In this chapter, we introduce *principal component analysis* (PCA) as another unsupervised learning method. PCA transforms the original data into another data representation where information is ordered in orthogonal dimensions. There are many applications, and in the following section, we focus on using PCA to reduce the number of data dimensions with information loss under control.

7.1 Dimension Reduction

When the dimensionality of the feature space is high, reducing this can help to find out the key structure of the original feature space. Suppose one tries to apply supervised learning methods like linear regression: working on data of lower dimensionality while keeping the most important information of the original dataset can be beneficial. There are two dimension reduction frameworks can be applied on the feature space:

- Feature selection: this selects a subset of features from the original variables. It can be done manually based on domain knowledge or using statistical tools.
- Feature projection: this transforms the high-dimensional data into lower-dimensional data by either linear or non-linear mapping of the data while preserving the most relevant information.

In the following sections, we will focus on principal component analysis (PCA), which is a feature projection method with a linear mapping.

7.2   Principal Component Analysis

The goal of PCA is to find an alternative representation of data X in a transformed space such that X can approximated by lower dimensional variables while preserving a given level of the original information.

Denote the data $X = (X_1, \ldots, X_p)$ as having p-dimensional features with zero mean; we aim to find a linear transformation V such that

- the p-dimensional transformed feature variable $Z := XV$ consists of components orthogonal to each other, and the variances of those components are in a descending order;
- the alternative representation of X: $X = ZV^{-1}$ can be seen as a projection onto the new feature space, based on which a dimension reduction approach can be built.

7.2.1   Linear transformation

Denote $V = (v_1, \ldots, v_p)$ as the loading matrix with $v_k = (v_{1k}, \ldots, v_{pk})^T$, and the linear transformation of X with V becomes

$$Z = XV. \tag{7.1}$$

PCA requires V to satisfy the following:

- Norm: Euclidean norm $||v_k||^2 = 1, \ \forall \ k \in \{1, \ldots, p\}$.
- Direction: With the norm specified, the direction of V is chosen such that the transformed variables $\{Z_k\}_{k=1}^p$ are all orthogonal to each other, and the variances are in a descending order: $var(Z_1) \geq var(Z_2) \geq \cdots \geq var(Z_p)$. We call Z_k the kth principal component for $k = 1, 2, \ldots, p$. It is also known as the PC scores.

The summary of finding the loading matrix V in PCA is in Algorithm 15. The algorithm provides a clear interpretation of the loading matrix, and in the next section we see how to solve the optimization problem involved.

7.2.2   Singular value decomposition

Singular value decomposition (SVD)[1] states that a real $N \times p$ dimensional matrix X can be decomposed as

$$X = UDV^T, \tag{7.2}$$

---

[1] Here we only provide the definition of SVD for the real matrix case.

Algorithm 15: PCA: Loading Matrix V

1: Find v_1:

$$v_1 = \underset{||v||=1}{\operatorname{argmax}} ||Xv||^2,$$

then $Z_1 := Xv_1$ has the largest variance among all possible projected variables, i.e., $||Z_1||^2 = \max_{v,||v||^2=1} ||Xv||^2$.

2: For $k = 2, \ldots, p$, find v_k:

$$v_k = \underset{||v||=1}{\operatorname{argmax}} ||\tilde{X}_k v||^2,$$

where $\tilde{X}_k$ is the residual of X after subtracting $k-1$ components reconstructed with PCs found from previous steps:

$$\tilde{X}_k = X - \sum_{j=1}^{k-1} \tilde{X}_j = X - \sum_{j=1}^{k-1} (Xv_j)\, v_j^T.$$

---

where

- U is an $N \times p$ orthonormal matrix: rows and columns of U are orthogonal, and $U^T U = I_p$ where I_p is the $p \times p$ dimensional identity matrix.
- V is a $p \times p$ orthonormal matrix: rows and columns of V are orthogonal, and $V^T V = I_p$.
- D is a $p \times p$ diagonal matrix with real entries $(d_i)_{i=1,\ldots,p}$ called singular values and $d_1 \geq d_2 \geq \ldots d_p \geq 0$.

Note that there are standard numerical methods to solve SVD to obtain U, D and V.

In this way the principal components Z can be obtained by

$$Z = XV = (UDV^T)V = UD. \tag{7.3}$$

7.2.3 Principal components and covariance

PCA is closely linked to the concept of covariance. From the analysis below we can see that if we assume all information is embedded in its covariance matrix, then PCA is able to extract information in an ordered way.

The empirical covariance matrix of X using SVD can be written as:

$$\Sigma_X = \frac{1}{N-1} X^T X = \frac{1}{N-1} (UDV^T)^T (UDV^T)$$

$$= V\left(\frac{1}{N-1} D^2\right) V^T. \tag{7.4}$$

Meanwhile, as the covariance matrix is a real non-negative symmetric matrix, it has the eigenvalue decomposition:

$$\Sigma_X = W\Lambda W^T, \tag{7.5}$$

where Λ is a diagonal matrix $\Lambda = diag(\lambda_1, \ldots, \lambda_p)$ with $\lambda_1, \ldots, \lambda_p$ being eigenvalues of Σ_X in descending order, and the columns of W are the corresponding eigenvectors. By comparing Equations (7.4) and (7.5), we can see that:

- $V = W$: the right singular vectors from SVD of X are the eigenvectors of the covariance matrix of X, i.e., the new coordinates V are provided by the eigenvectors of the covariance matrix of the original system.
- $\lambda_j = \frac{d_j^2}{N-1}$: the eigenvalue of the covariance matrix of X is the scaled square of singular values from the SVD of X.

Moreover, we can see that the empirical covariance matrix of the principal components Z is

$$\Sigma_Z = \frac{1}{N-1}Z^T Z = \frac{1}{N-1}(XV)^T XV = V^T \Sigma_X V = \frac{D^2}{N-1} = \Lambda.$$

Therefore we can conclude:

- The principal components $Z_1, \ldots, Z_p$ are orthogonal, and the variances of $Z_1, \ldots, Z_p$ are the eigenvalues of the covariance matrix of X in descending order.
- With the representation of the covariance of the data using SVD, we can apply PCA to reduce the dimensionality of data.

7.2.4 Introduction

Once we obtain the principal components and the loading matrix, the representation of X can be written as

$$X = ZV^{-1} = ZV^T = \sum_{j=1}^{p} Z_j v_j^T. \tag{7.6}$$

Therefore we obtain an approximating sequence $\left\{ \tilde{X}^{(k)} \right\}_{k=1,\ldots,p}$ of feature variables X, with $\tilde{X}^{(k)}$ including the first k principal components of X:

$$\tilde{X}^{(k)} = \sum_{j=1}^{k} Z_j v_j^T. \tag{7.7}$$

The residual from the approximation using k PCs is

$$R^{(k)} = X - \tilde{X}^{(k)} = \sum_{j=k+1}^{p} Z_j v_j^T. \qquad (7.8)$$

- When $k = p$, $\tilde{X}^{(p)}$ recovers the original data X.
- From the variance analysis in Section 7.2.3, $\tilde{X}^{(k)}$ explains $\frac{\sum_{j=1}^{k} \lambda_j}{\sum_{j=1}^{p} \lambda_j}$ of the variance of X.

Therefore, assuming the information is embedded in the covariance, given a dataset X and an information keeping criterion in terms of percentage of variance explained, we can find an approximation of X with reduced dimensionality.

7.2.5 Practical problems

Before we discuss an example of using PCA in Section 7.4, we briefly overview a few problems in practice below.

Scaled or not—Note that PCA is not invariant to scaling, so we need to consider whether to standardize the data input X or not, in addition to removing the mean from the data, before applying PCA. This depends on the context of the data and the objective, and the following are some tips:

- If features are measured in different units that are not comparable, we should first standardize them before applying PCA.
- If features are measured in comparable units, then we can keep them unscaled to preserve the original variability information.

What if the data are not Gaussian? PCA is based on the assumption that the information is entirely represented by the covariance. As a Gaussian distribution—or more broadly the spherical distribution family—can be fully characterized by the first two moments, applying PCA on data generated from a Gaussian distribution would give reasonable and effective results. However, if the data X are generated from a non-Gaussian distribution, then the covariance is not sufficient to characterize all the information embedded. In this case, we can use alternative methods such as *independent component analysis* (ICA).

7.3　Python Implementation

The usage of PCA functions in the Scikit-Learn package is shown in Listing 7.24.

```python
import numpy as np
import pandas as pd
from sklearn.decomposition import PCA

class PCABase(object):
    def __init__(self,X):
        self.X = X
        self.n_features = X.shape[1]
        self.dates = X.index
        self.Xc = self.X - self.X.mean() # centered
        self.pc_names = lambda n: ['PC'+str(i) for i in np.arange(1,n+1)]

    def pca(self,n_pc=None):
        # fit pca model
        if n_pc:
            model = PCA(n_components=n_pc).fit(self.Xc)
        else:
            model = PCA().fit(self.Xc)
        return model

    def cps(self):
        # loading matrix => principal axes in feature space
        cps = self.pca().components_.T
        return self.to_df_pc(cps,is_loading=True)

    def cumsum_expvar_ratio(self):
        var_exp = self.pca().explained_variance_ratio_
        var_exp_cumsum = np.cumsum(var_exp)
        return var_exp,var_exp_cumsum

    def scores(self):
        scores = self.pca().transform(self.Xc)
        return self.to_df_pc(scores)

    def to_df_pc(self,data,is_loading=False):
        cols = self.pc_names(self.n_features)
        idx = self.X.columns if is_loading else self.dates
        return pd.DataFrame(data,columns=cols,index=idx)
```

Listing 7.24. PCA: Base class.

7.4 Application: Term Structure Analysis Using PCA

7.4.1 Introduction to fixed income term structure

There are many financial instruments available in the market with maturity date specified in the contract. Examples are bond products issued by governments or corporations, future, swaps and options. Bonds issued by a government can have maturities from months to over 30 years. We say the bond yield has a term structure as a yield level depends on the underlying bond maturity.

Different types of investors enter into the contracts with different motivations. Some investors need to hedge their existing positions, and some investors aim to make money with a view about how the market will move in the future. An insurance company collects insurance premiums on the one hand and pays out claims on the other, so it usually invests heavily in the bond market to preserve a balance between assets and liabilities. A hedge fund also invests in such a market once a portfolio manager finds a market dislocation based on their market view. For instance, if she finds the 2Y bond is relatively cheaper while the 10Y bond is more expensive compared to normal levels, and she believes both will revert to the normal level, then she would buy 2Y and sell 10Y now.

Regardless of the motivation, to be able to invest at the right time, all investors need a good understanding about how the yield curve moves.

Denote by X the yield curve consisting of bonds of p maturities with N observations. The most common framework for modeling X is to use a k factor model.

Factor Model

$$X = \begin{pmatrix} x_1 \\ \vdots \\ x_N \end{pmatrix} = \begin{pmatrix} x_{1,1}, & x_{1,2}, & \cdots, & x_{1,p} \\ \vdots & \vdots & & \vdots \\ x_{N,1}, & x_{N,2}, & \cdots, & x_{N,p} \end{pmatrix}$$

The k-dimensional linear factor model for X can be written as

$$X = \mu_X + Zf + e, \tag{7.9}$$

where $\mu_X = (\mu_1, \ldots, \mu_p)$ is the mean vector of X, $e = (e_1, \ldots, e_N)$ is the residuals and

$$Z = \begin{pmatrix} z_{1,1}, & z_{1,2}, & \cdots, & z_{1,k} \\ \vdots & \vdots & & \vdots \\ z_{N,1}, & z_{N,2}, & \cdots, & z_{N,k} \end{pmatrix} \text{ consists of } k \text{ factors,}$$

$$f = \begin{pmatrix} f_{1,1}, & \cdots, & f_{1,p} \\ \vdots & & \vdots \\ f_{k,1}, & \cdots, & f_{k,p} \end{pmatrix} \text{ is the factor loading matrix;}$$

i.e., the observation i of feature j is

$$x_{i,j} = \mu_j + \sum_{l=1}^{k} z_{i,l} f_{l,j} + e_{i,j}. \tag{7.10}$$

The question becomes how to determine the factors.

One can define factors such as macroeconomic factors like GDP, unemployment rate, FX rate and so on. These have certain explanatory power and may be able to explain the curve change pattern over very long period like five years, but they may not be sufficient to explain the yield curve due to the following reasons:

- we have less frequent data observable on key economic factors;
- a highly correlated set of macroeconomic factors may not be able to provide a stable result from the fitting process.

Instead of specifying external factors beforehand, we could apply PCA to:

- explore the self-contained independent driving factors and understand the key structure of the yield curve dynamic; we will be able to answer how 2Y, 10Y and 30Y bond yields move together using those factors;
- reduce the factor dimensionality naturally based on the explained variance ratio;
- focus on hedging based on risk exposure representation in terms of the factors.

7.4.2   Data and observation

The top figure in Figure 7.1 shows the time series of yield/rate of different maturities, and we can see that the rates seem to move in a similar way, indicating that some common factors drive the yield dynamics. The bottom

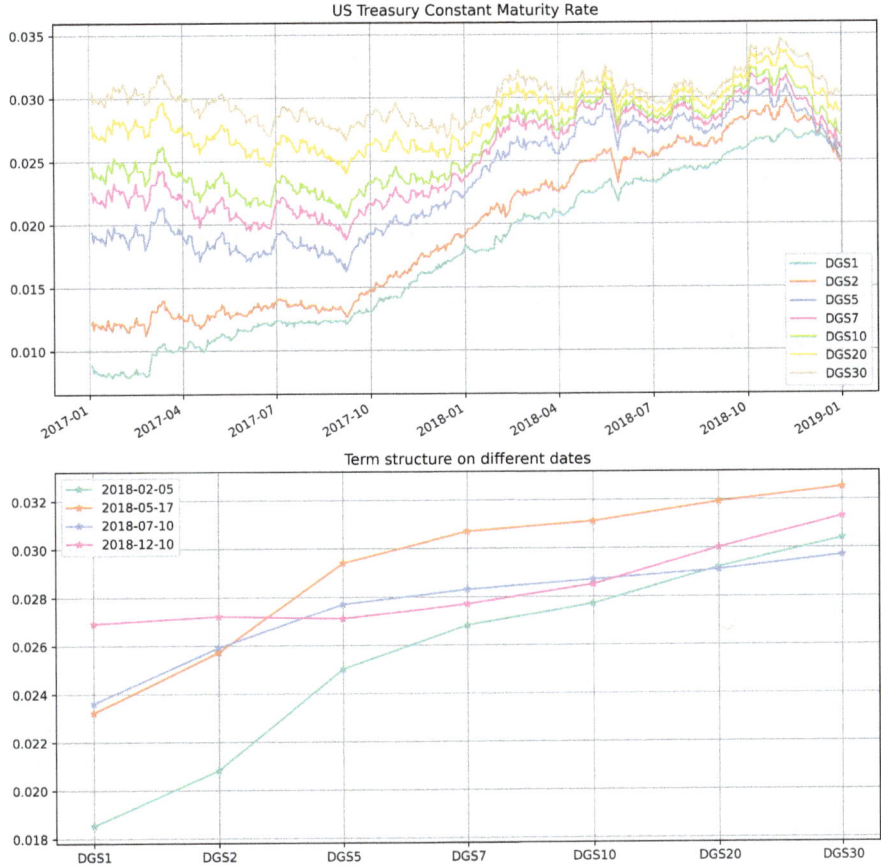

Figure 7.1. US Treasury constant maturity rates time series and term structure.

figure in Figure 7.1 shows the yield curve term structure on different dates, and we can see a few different types of changes between dates:

(1) Level change: For example, between 2018/02/05 and 2018/05/17, the main difference is the level, and the shape did not change much.

(2) Slope change: From 2018/05/17 to 2018/12/20, the slope of the curve changed as the short-term yield increased while the long-term yield decreased.

(3) Curvature change: besides the level and slope change, the curve's curvature also changes between dates. The curvature change reflects how the difference between the mid-long premium (i.e., long-term minus mid-term rate level) and the short-mid premium (i.e., mid-term minus

short-term rate level) is changing from day to day. In other words, the curve "belly" moves relative to short-term and long-term level changes daily.

7.4.3 PCA on term structure

Since we have seen from previous sections that the yield curve dynamic seems to be driven by some common factors, PCA may help us to explore those factors. Consider X as the de-meaned daily yield change of p maturities on N days; from Equation 7.3 in Section 7.2, we obtain the representation of X as $X = ZV^T$ in the PCA framework. We analyze the PCA results in the following sections.

7.4.3.1 *Principal components*

Z consists of the principal components. Figure 7.2 shows the time series of the first three principal components, where we can see the variance of the PCs are in descending order. It also can be seen from Figure 7.3 that PC1 explains more than 80% of the total data variance, and the first three PCs explain almost 100% of the variance. This indicates that the whole yield curve dynamic can be explained by the first three PCs. Next we will explore how those PCs represent the dynamic from the loading matrix.

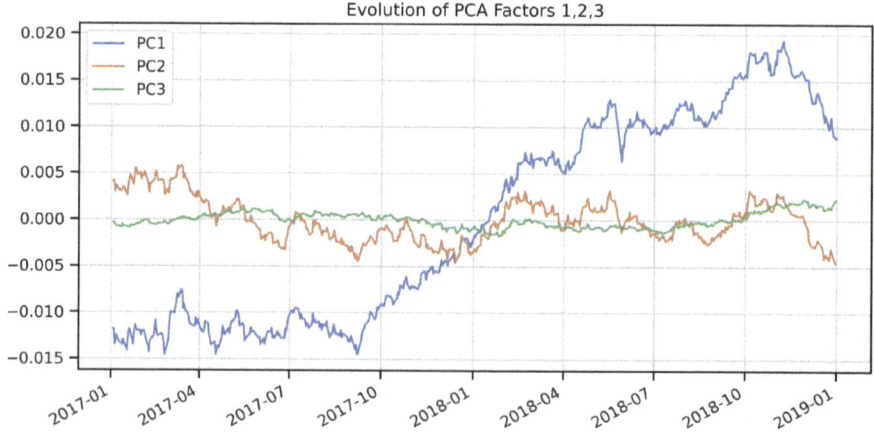

Figure 7.2. Principal components.

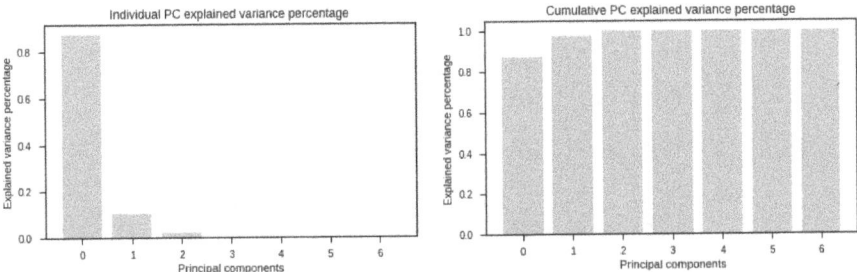

Figure 7.3. PCA-explained variance ratio.

7.4.3.2 *Loading matrix*

V is loading matrix representing X in the new coordinate system, which can be found in the top table of Figure 7.4. The visualization of the PCA loading matrix is given in the bottom part of Figure 7.4.

PC1: The observed feature about PC1 is that the loadings are all positive; therefore the factor PC1 leads the term structure to move in the same direction. It represents the level change of the curve. In addition, it reflects the reality that the short-term yield levels tend to move more than long-term ones as loadings on the short end are larger. In addition, from Figure 7.3 we see that most of the variance of the curve move comes from this factor.

PC2: The observed feature about PC2 is that the loading increases from a negative value on the short-end to positive values on the long-end, and moreover it crosses zero once between 2Y and 5Y. This indicates the short-term yield and long-term yield move in a different direction, thus it characterizes the slope of the curve. Moreover, from the zero-crossing point, we can see that the anchor point on the curve slope move is around 5Y.

PC3: The observed feature about PC3 is that the loadings from short- to long-term cross zero twice. It characterizes the fact that the very short-term and long-term yields move in one direction while the belly part of the curve tends to move in the opposite way. This PC naturally reflects the curvature structure of the curve.

As we see from above that the first three PCs seem to coincide with the level, slope and curvature factors observable from the historical data, we can verify this finding. We choose 10Y yield for the level, 10Y − 2Y as the

	PC1	PC2	PC3	PC4	PC5	PC6	PC7
PC1	1.00	0.42	0.20	0.53	-0.44	-0.33	-0.25
PC2	0.42	1.00	-0.72	-0.27	0.31	0.20	-0.21
PC3	0.20	-0.72	1.00	0.61	-0.46	-0.43	0.08
PC4	0.53	-0.27	0.61	1.00	-0.53	-0.31	-0.20
PC5	-0.44	0.31	-0.46	-0.53	1.00	0.48	0.01
PC6	-0.33	0.20	-0.43	-0.31	0.48	1.00	-0.02
PC7	-0.25	-0.21	0.08	-0.20	0.01	-0.02	1.00

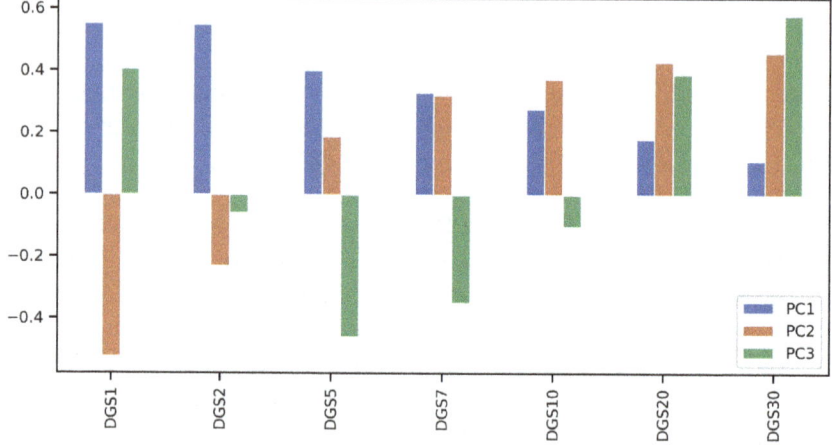

Figure 7.4. PCA loading matrix.

slope and $5Y - 2 \times 10Y + 30Y$ as the curvature. We compare these with the first three PCs in Figure 7.5. We do observe similar patterns.

7.4.3.3 *Representation of X with PCs*

Once the underlying factors have been found from PCA, here we see how well the factors explain the dynamics of the original data. Since most of the variance can be explained by the first three PCs as seen from Figure 7.3, we use those PCs to reconstruct $\tilde{X}^{(3)}$ as an approximation of the original X via Equation 7.7, and the corresponding residual is $R^{(3)}$ via Equation 7.8. The top plots in Figures 7.6 and 7.7 show the reconstructed 10Y and 30Y curves, respectively, and the bottom plots are the corresponding residuals.

- 10Y: PC1 dominates the explanatory power; including PC2 improves this a little while including PC3 in addition does not make a noticeable

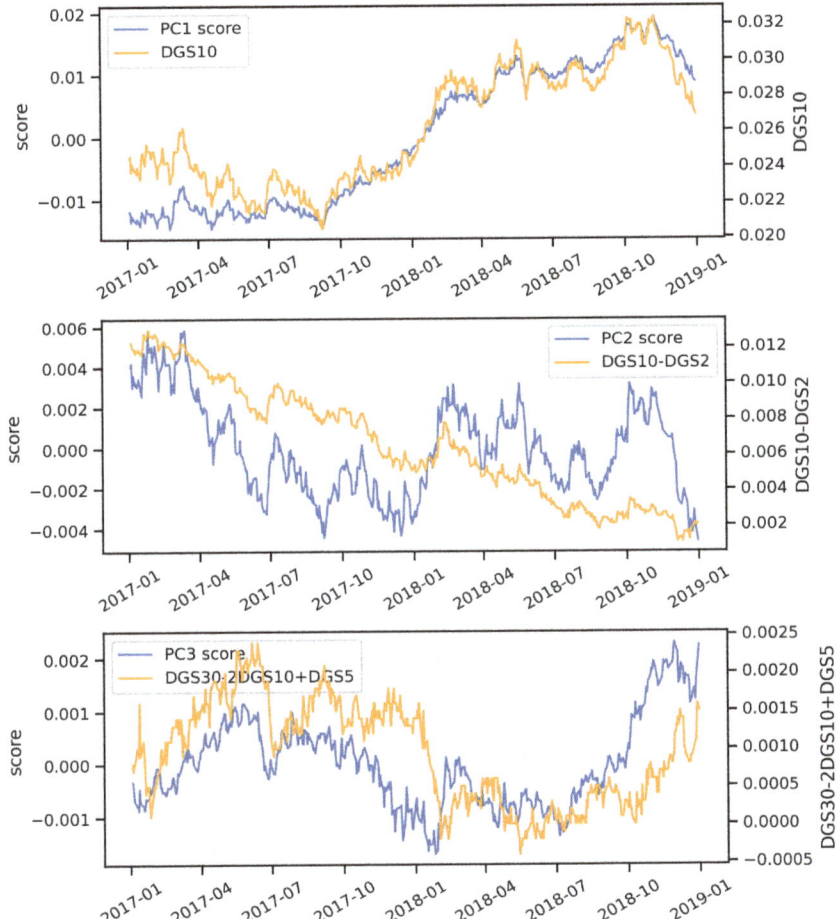

Figure 7.5. What do PC scores represent?

difference. This means that PC1 and PC2 are sufficient to explain the dynamics.

- 30Y: Compared to the 10Y curve, PC2 plays a more significant role in explaining the 30Y move. The explanatory power from PC3 is also more than for 10Y.

7.4.4 PCA for hedging

After exploring the principal components and their interpretation, in this section we focus on using PCA to hedge the portfolio risk.

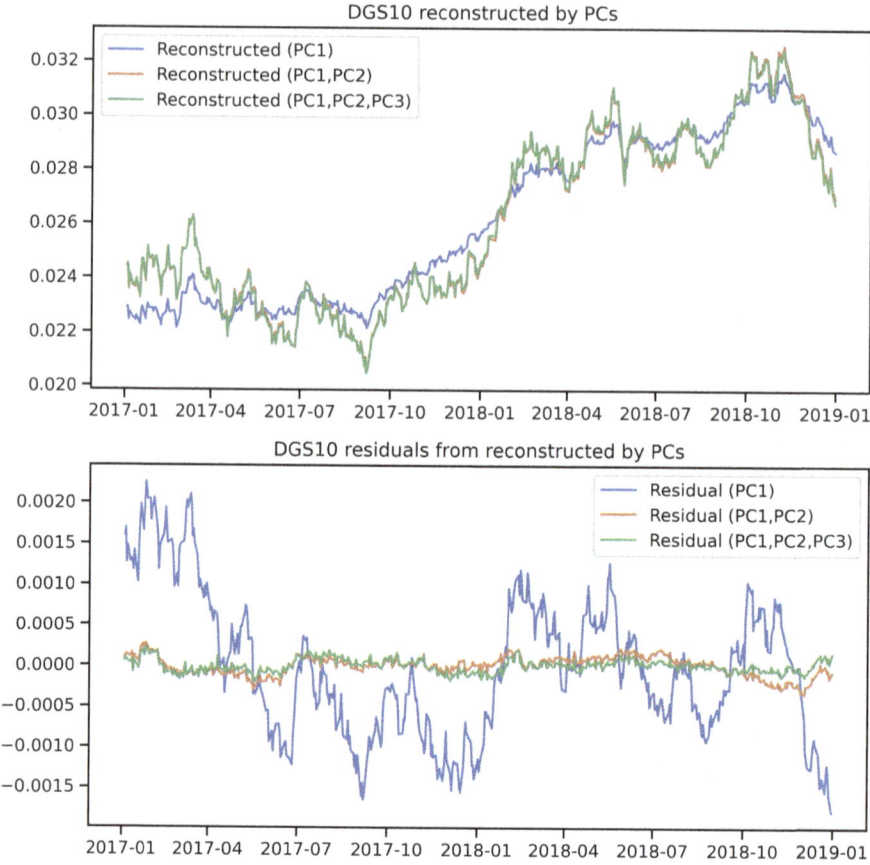

Figure 7.6. Residuals of the reconstructed series from PC: DGS10Y.

Risk comes together with investment. Why is hedging important? Consider two examples below:

- If you hold a portfolio of bonds, and you want to hedge all risks coming from the yield curve move regularly.
- If you have a view on how the slope of the yield curve will move next week after a major economic event though you are not sure about how much the overall level would change. Therefore you are motivated to position yourself in a slope trade—e.g., buy 10Y sell 2Y if you expect the slope of 2Y and 10Y is going to be flatter than the current curve. Meanwhile you also need to hedge your level risk to be neutral on the level change.

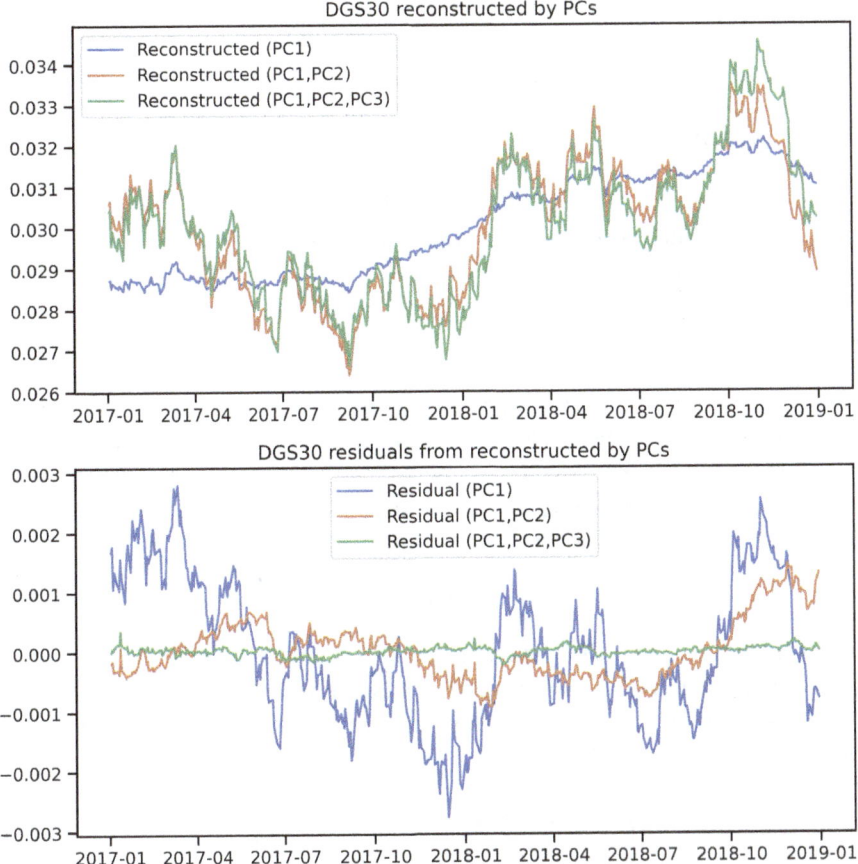

Figure 7.7. Residuals of the reconstructed series from PC: DGS30Y.

To be more general, no matter if you are a passive or active investor, the following need to be considered to make a hedging decision:

- What constitutes the current position risk? Can we quantify it?
  If the yield curve moves, the value of the bond portfolio will change accordingly. We have seen that the risk exposure curve can be decomposed into level, slope and curvature.
- What types of risks are you willing to take, and how much tolerance do you have? How often do you want to hedge?
  This depends on the investment mandate or the trading strategy. One may need to hedge more often when the macro environment is changing rapidly.

- Which instruments do you choose for hedging?
  Once the risk profile target has been decided, we need to choose the instruments used to hedge as different instruments have different risk exposures. Liquid products are good candidates for hedging instruments in general.

As seen from previous sections, PCA is able to describe the whole curve dynamic in terms of level, slope and curvature; hence, it can be used to hedge the portfolio effectively.

7.4.4.1 *Risk representation*

Consider a yield term structure consisting of bonds with p maturities. At time t, denote the yield of maturity T_m as r_t^m and the price of the corresponding bond as P_t^m. Then the whole yield term structure is $r_t = (r_t^1, ..., r_t^M)$, and its daily change dr of N observations is an $N \times p$ matrix with the following PC representation:

$$dr = (UD)V^T = ZV^T, \qquad (7.11)$$

where Z is the $N \times p$ PCs and V is the $p \times p$ loading matrix.

We call the portfolio risk from the yield curve move the **delta risk**. The delta risk of bond m with respect to maturity k is $\delta_t^{m,k} := \frac{\partial P_t^m}{\partial r_{t,k}}$, thus the price change can be written as

$$dP_t^m = \sum_{k=1}^m \frac{\partial P_t^m}{\partial r_{t,k}} dr_{t,k} = \sum_{k=1}^m \delta_t^{m,k} dr_{t,k}. \qquad (7.12)$$

Now we express the price change in terms of sensitivity to the PCs instead of to the original yield:

$$dP_t^m = \sum_{j=1}^p \frac{\partial P_t^m}{\partial Z_{t,j}} dZ_{t,j}, \qquad (7.13)$$

where $\frac{\partial P_t^m}{\partial Z_{t,j}}$ is jth PC risk exposure representing the sensitivity of the price of bond m to the jth PC. From Equation (7.11), we have

$$\frac{\partial P_t^m}{\partial Z_{t,j}} = \sum_{k=1}^m \frac{\partial P_t^m}{\partial r_{t,k}} \frac{\partial r_{t,k}}{\partial Z_{t,j}} = \sum_{k=1}^m \delta_t^{m,k} v_{k,j}. \qquad (7.14)$$

Therefore, assuming the price change can be largely explained by the first three principal components, from Equations (7.13) and (7.14),

$$dP_t^m = \sum_{j=1}^{p} \left(\sum_{k=1}^{m} \delta_t^{m,k} v_{m,j} \right) dZ_{t,j}. \tag{7.15}$$

Equations (7.12) and (7.15) provide two ways to represent risk in the original instrument and the PCs, respectively.

7.4.4.2 *Hedge level with PCA*

Suppose we choose bond T_k to hedge the exposure to the first PC; then we can solve the hedge quantity $w_{t,1}^k$ by:

$$\sum_{m=1}^{p} \frac{\partial P_t^m}{\partial Z_{t,1}} = w_{t,1}^k \frac{\partial P_t^k}{\partial Z_{t,1}}. \tag{7.16}$$

7.4.4.3 *Hedge level and slope with PCA*

We need to use two instruments to hedge both level and slope risks represented by the first and second PCs. Assume we hedge with instruments k_1 and k_2; then we can work out the quantity $(w_{t,2}^{k_1}, w_{t,2}^{k_2})$ by solving the following equation system:

$$\sum_{m=1}^{p} \frac{\partial P_t^m}{\partial Z_{t,1}} = w_{t,2}^{k_1} \frac{\partial P_t^{k_1}}{\partial Z_{t,1}} + w_{t,2}^{k_2} \frac{\partial P_t^{k_2}}{\partial Z_{t,1}}, \tag{7.17}$$

$$\sum_{m=1}^{p} \frac{\partial P_t^m}{\partial Z_{t,2}} = w_{t,2}^{k_1} \frac{\partial P_t^{k_1}}{\partial Z_{t,2}} + w_{t,2}^{k_2} \frac{\partial P_t^{k_2}}{\partial Z_{t,2}}. \tag{7.18}$$

A similar method can be applied to hedge higher order PCs.

7.4.4.4 *Comments*

- Relative Value Analysis with PCA
  Equation 7.8 provides residuals after removing the components from the first k PCs. One can build trading strategies exploring opportunities from slope or curvature move of the curve by analyzing the features of the residual time series. For instance, the residual after removing the first two PCs is typically a mean reverting process, if the current residuals deviate from the mean without big changes in the market

environment, then one can build a curve position on the curvature to take advantage of this.

- Can the curve always be explained by the level, slope and curvature from PCA? One can find an interesting discussion about this by Lord and Pelsser [Lord and Pelsser (2007)].

7.4.5 Clustering and PCA

One may encounter the question of whether there needs to be data segmentation when applying PCA. Time series data segmentation indicates that there is a regime switch embedded in the data. Therefore the PCA should be applied to each subperiod's data samples instead of to the whole, otherwise the results will not explain the data well. There are many ways to detect a regime switch or the change point. Here we show that we can use the clustering methods introduced in Chapter 6 for the initial exploration.

Take the term structure application in Section 7.4.3 as an example. We apply the K-means clustering method to detect the data segmentation. Instead of looking into each maturity, we first extract the PC1, PC2 scores obtained from the entire period 2017/01/03 to 2018/12/31 as features since they contain the most variance information in the data. After performing K-means with $K = 2$, from the clustering results in Figure 7.8, we can see that the time series can be broken down into two clusters: one is year

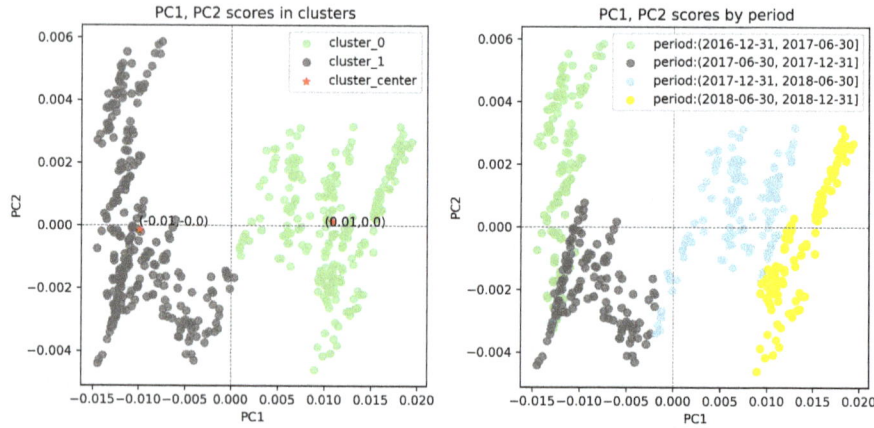

Figure 7.8. The first two PC scores from the original time series. Left: scatter plot of PC1 and PC2 scores clustered by two-cluster K-means. Right: PC1, PC2 scores indexed by 6M period bucket.

2017 and the other is 2018. Considering the subperiod data sample size, and also observing the original data in Figure 7.1, using the 1-year data window seems to provide more effective results. Moreover, recall that PC1 and PC2 characterize the curve level and slope, which we have seen in previous sections; the clustering results indicate that we have two regimes if we consider level and slope changes.

From this example we show that applying clustering on features extracted from PCA can provide information not only about the data segmentation on the timeline but also about interpretable structure change dimensions.

7.5   Exercises

(1) Explain the assumptions behind PCA.
(2) What is SVD?
(3) Explain the loading matrix of PCA.

Chapter 8

Reinforcement Learning

8.1 Introduction

Reinforcement learning (RL) is one of the hottest topics in the machine learning field as it has applications in various areas. The success of AlphaGo underpins the enormous potential of reinforcement learning.

RL is an area of machine learning that aims to solve how an *agent* ought to take *actions* in an *environment* to maximize some notion of cumulative *reward*. It mimics the way of humans to learn and make decisions by interacting with our environment based on experience, which uses rewards and punishment as signals for positive and negative behavior with the ultimate goal of finding optimal action. Whether we are learning to plan a trip or to start a conversation, we take action based on a current situation. How the environment responds to our action shapes our perception of the action and affects what we do next in return. We seek to influence what happens through our behavior. RL, the approach we explore, is much more focused on goal-directed learning from interaction than are the other approaches we have discussed previously. Important terminologies of reinforcement learning and their descriptions are listed in Table 8.1.[1]

An RL problem can be well explained through trading as an example. Let us consider trading assets where the goal of the investor is to maximize the final wealth for her trading strategy. A stock market composed of price process of a set of stocks is an interactive environment for the investor. Her wealth is a reward of using this trading strategy. States are admissible trading strategies (e.g., self-financing), and the total cumulative reward is the final wealth. In RL, the optimal action is learned by interacting with

---

[1] `https://www.kdnuggets.com/2018/03/5-things-reinforcement-learning.html` provides a concise overview of RL.

Table 8.1. Key terminologies of reinforcement learning.

- Environment: Physical world in which the agent operates.
- State: Current situation of the agent.
- Reward: Feedback from the environment.
- Policy: Method to map state space to action space.
- Value: The expected future reward that an agent would receive by taking an action in a particular state.

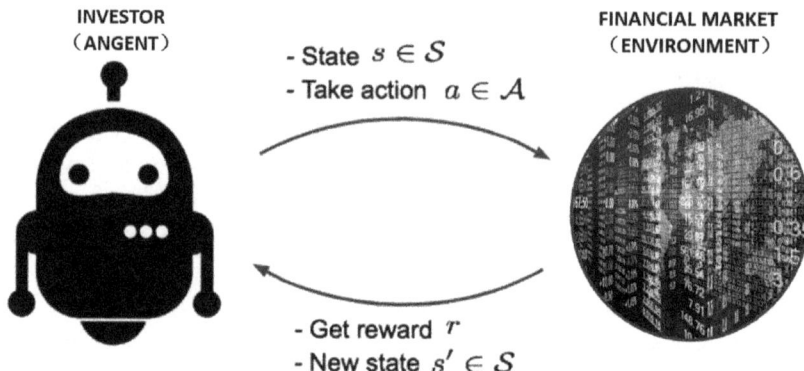

Figure 8.1. Illustration of RL in the trading context.

the market (environment) and learning from the performance of varying trading strategies given the environment. Figure 8.1 shows an illustration of RL in the trading example.

Mathematically, a reinforcement learning problem can be formulated as a stochastic optimal control problem. Dynamic programming (DP) is a classical solution method for the stochastic optimal control problem that elevates it to a more general time-indexed optimization problem, solves this more generally-formulated problem via a difference equation or partial differential equation (Bellman's equation), and projects the solution to the starting time to solve the overall problem. In DP, the environment is typically formulated as a Markov Decision Process (MDP) of specific form. Compared with classical dynamic programming methods, reinforcement learning algorithms have much less model assumptions than the MDP and they target a general class of MDP where exact methods become infeasible.

DP suffers from the curse of dimensionality, meaning that its computational requirements grow exponentially with the number of state variables. Still, it has had successful applications in diverse areas, such as bioinformatics, finance and engineering. Compared with DP, reinforcement learning also uses MDP to formulate an environment. However, it allows a model-free environment and utilizes DP techniques to iteratively update policy to eventually build optimal policy instead of finding an analytic solution for the optimal policy as in the classical DP setting. To build an optimal strategy, the agent has to choose between exploring new states and maximizing its reward based on experience. This is called the Exploration vs. Exploitation trade-off.[2]

There are several popular reinforcement algorithms. Depending on whether the RL algorithm relies on the approximation of value functions, RL can be divided into the following two types:

(1) Direct reinforcement (DR) learning (learning actions, actor-based): DR defines a spectrum of continuous actions directly from a parameterized family of policies. The optimization of actor-based learning is much simpler than the indirect reinforcement learning discussed below, as DL only requires a differentiable objective function with latent parameters [Deng *et al.* (2017)]. There is usually no need to approximate and evaluate the value function to obtain the optimal policy/action.

(2) Indirect reinforcement learning (learning value function, critic-based): In a typical indirect method, the optimization always relies on complicated dynamic programming methods to derive optimal actions on each state, e.g., Q-Learning and TD-Learning (temporal difference learning).

In this chapter, we focus on how to apply RL to design the optimal trading strategy to achieve the best investment objective. The optimal trading problem can be formulated either as a one-stage or a two-stage problem. The one stage problem is usually in the form of the stochastic optimal control problem to learn an optimal strategy in an end-to-end manner. In contrast, a two-stage problem subdivides the optimal trading into two sub-problems: (1) the forecasting problem of future price movements and (2) the question of how to build an optimal trading strategy based on the price movement predictor. Here we focus on the one-stage problem setting as we would like to use RL to approach this problem. Following

---

[2]Interested readers can refer to [Sutton and Barto (2018)] for more information.

[Moody and Saffell (2001)] and [Deng *et al.* (2017)], we choose to use direct learning (actor-based method), because direct learning exhibits two advantages as follows:

- flexible objective functions for optimization (one has the flexibility to choose different objective functions, e.g., terminal wealth or Sharpe ratio);
- continuous descriptions of market conditions (instead of discrete states as in indirect learning).

Therefore, direct learning may provide a better framework for trading than Q-learning or other similar indirect approaches. In this chapter, we present one type of direct reinforcement learning, called *recurrent reinforcement learning* (RRL), to tackle the problem of portfolio optimization proposed in [Moody and Saffell (2001)]. RRL provides a simpler problem representation, which avoids the curse of dimensionality and offers compelling advantages in terms of efficiency. In extensive work on both synthetic data and real financial data, it has been shown that the approach based on RRL produces better trading strategies than systems utilizing Q-Learning [Moody and Saffell (2001)].

8.2   Recurrent Reinforcement Learning

Recurrent reinforcement learning (RRL) is an adaptive approach of direct learning for portfolio optimization, which is proposed by John Moody and Matthew Saffell in [Moody and Saffell (2001)]. For simplicity, we only consider a single risky asset case instead of a portfolio. Let us start with the explanation of the structure of single asset trading systems and the setup of the problem. For ease of discussion, the following notations are introduced:

- z_t is the price of the asset at time t;
- r_t is the return at time t, i.e., $r_t = z_t - z_{t-1}$[3];
- r_t^f is the risk-free return at time t;
- F_t is the trading position of the asset at time t, where

$$F_t = F(\theta_t; F_{t-1}, I_t) \text{ with}$$

$$I_t = \{z_t, z_{t-1}, \cdots ; y_t, y_{t-1}, \cdots\},$$

and y_t denotes other external variables of arbitrary number;

---

[3]It represents the additive profit, which is appropriate for trading a fixed amount of shares or contracts of a security such as futures or FX contracts. For multiplicative profits, i.e., $r_t = z_t/z_{t-1} - 1$, the formulation of R_t would be slightly different.

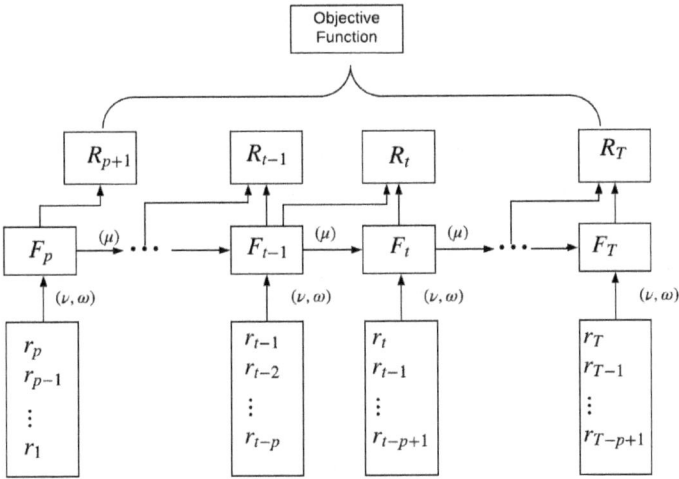

Figure 8.2. The graphical illustration of RRL algorithm.

- R_t is the trading return at time t,

$$R_t \equiv u\{r_t^f + F_{t-1}(r_t - r_t^f) - \delta|F_t - F_{t-1}|\},$$

where $u > 0$ is the trading position size.

Figure 8.2 illustrates the structure of the above variables in the RRL algorithm. The goal is to learn the optimal parameters in the model in order to maximize the Sharpe ratio of the trading strategy, which is denoted by S_T and defined as

$$S_T = \frac{\text{Average}(R_t)}{\text{Standard Deviation}(R_t)}.$$

Here we consider the only possible trades of taking long or short positions, i.e., $F_t \in \{1, -1\}$. Assuming that the current trading strategy F_t depends on the previous trading strategy F_{t-1} and p-lagged return values, we adopt the following simple model for F_t:

$$F_t = \text{sign}\left(\mu F_{t-1} + \sum_{i=0}^{p-1} \nu_i r_{t-i} + \omega\right),$$

where $F_t \in \{1, -1\}$,[4] and $\theta \equiv (\mu, \nu_0 \cdots, \nu_{p-1}, \omega)$ are model parameters.

---

[4] Since the sign is not differentiable, we replace it with the differentiable activation function tanh during learning and discretize the results when trading.

Let x_t denote the p-lagged value of the return, i.e., $x_t = (r_{t-p+1}, \cdots, r_t)^T$. Then $(F_t)_{t=p}^{T-1}$ can be regarded as the hidden layer of the recurrent network with $(x_t)_{t=p}^{T-1}$ as the input layer and sign as the activation function.

The next step is to utilize the numerical optimization scheme to estimate the optimal parameters in the model. Here we choose gradient descent for the optimization, i.e.,

$$\Delta\theta = \rho \frac{dS_T(\theta)}{d\theta}.$$

By the chain rule, we can get

$$\frac{dS_T(\theta)}{d\theta} = \sum_{i=1}^{T} \frac{dS_T}{dR_i} \frac{dR_i}{d\theta} = \sum_{i=1}^{T} \frac{dS_T}{dR_i} \left\{ \frac{dR_i}{dF_i} \frac{dF_i}{d\theta} + \frac{dR_i}{dF_{i-1}} \frac{dF_{i-1}}{d\theta} \right\}.$$

It is seen that $\frac{dS_T(\theta)}{d\theta}$ depends on the whole history of $(R)_{i=1}^{T}$ due to the recurrence structure of F_t. Therefore the key thing is how we compute the gradient $\frac{dS_T(\theta)}{d\theta}$ efficiently. This reminds us of the TBPTT algorithm for RNN we discussed in Section 5.4.5.

8.3 Link between RRL and RNN

In Chapter 5, we showed how to use Keras to implement a neural network model for supervised learning problems in a neat and concise way. This motivates us to reformulate the RRL algorithm as a supervised learning problem with a recurrent neural unit as the building block.

Let us design a neural network architecture that takes $(I_t)_{t=p+1}^{T}$, $(R_t)_{t=p}^{T-1}$ as the input and output layer.

(1) **Input Layer:** $(I_t)_{t=p+1}^{T}$ where $I_t = \begin{pmatrix} x_t \\ r_{t+1} \end{pmatrix}$.

(2) **Hidden Layer:** $(Y_t)_{t=p}^{T}$ where $Y_t = \begin{pmatrix} F_t \\ r_{t+1} \end{pmatrix}$.

It is constructed by the following three steps:

  Step 1: We split the input layer into two parts: $(X_t)_{t=p}^{T-1}$ and $(r_{t+1})_{t=p}^{T-1}$.

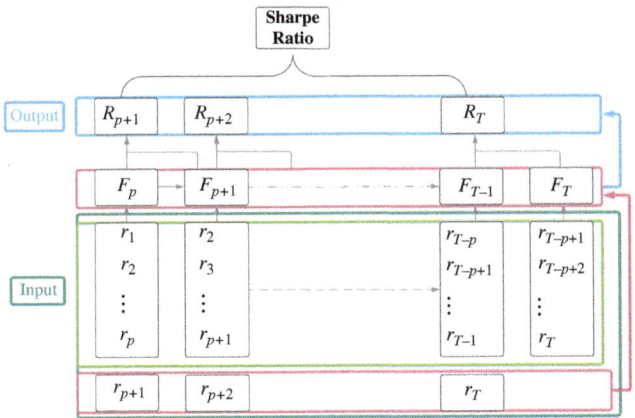

Figure 8.3. Reformulation of RRL algorithm as a neural network.

Step 2: Set up the RNN Layer: $(X_t)_{t=p}^{T} \to (F_t)_{t=p}^{T}$.

Step 3: Merge $(F_t)_{t=p}^{T}$ and $(r_{t+1})_{t=p}^{T-1}$ to obtain $(Y_t)_{t=p}^{T}$.[5]

(3) **Output Layer:** $(R_{t+1})_{t=p}^{T-1}$, which can be directly computed from $(Y_t)_{t=p}^{T-1}$ based on

$$R_t := u\{r_t^f + F_{t-1}(r_t - r_t^f) - \delta|F_t - F_{t-1}|\},$$

as $Y_t = (F_t, r_{t+1})_{t=p}^{T}$ for the given r_t^f, u and δ.

Figure 8.3 gives a summary of the neural network built as above, where the dark green block represents the input layer, the pink one represents the hidden layer composed of $(F_{t-1}, r_t)_{t=p+1}^{T}$ and the blue block denotes the output layer representing the trading return series. The customized intermediate layers to construct the desired output layer of the RRL model are given in Listing 8.25.

The negative Sharpe ratio of the output layer is chosen for the loss function in this case. In terms of the implementation using Keras in Python, we need to write a customized loss function (Listing 8.26), which is different to the standard loss function of the additive form, such as mean squared error. Python code to construct the neural network model for RRL is given in Listing 8.27.

---

[5]To merge $(F_t)_{t=p}^{T}$ and $(r_{t+1})_{t=p}^{T-1}$ of different length, we set $r_{T+1} = 0$.

```
1   # Import keras and tensorflow modules
2   import tensorflow as tf
3   from tensorflow import keras
4   from tensorflow.keras.models import Model, Sequential
5   from tensorflow.keras.layers import SimpleRNN, Dot, Dense, Activation,
    ↪   Input, Lambda, Add, Flatten, Multiply, Concatenate, Subtract, Layer
6   from tensorflow.keras import optimizers
7   import tensorflow.keras.backend as K
8
9   tf.random.set_seed(1)   # Set the random seed
10
11  # Define lambda function to split.
12  def split_func(x, p, flag):
13    split1, split2 = tf.split(x, [p, 1], -1)
14    return split1 if flag == 1 else split2
15
16  def trading_return(x, delta, T):
17    F_t_Layer, r_tplus1_Layer = tf.split(x, [1, 1], -1)
18    F_tminus1_layer1, f1 = tf.split(F_t_Layer, [T, 1], -2)
19    f1, F_t_layer1 = tf.split(F_t_Layer, [1, T], -2)
20    transaction_part = delta*tf.abs(tf.subtract(F_t_layer1,
    ↪   F_tminus1_layer1))
21    return_profit_layer  = Multiply()([F_t_Layer, r_tplus1_Layer])
22    f2, output_layer = tf.split(return_profit_layer, [1, T], -2)
23    output_layer = output_layer-transaction_part
24    return output_layer
```

Listing 8.25. Define the customized layers to construct the desired intermediate layers for the RRL models.

```
1   def sharpe_ratio_loss(yTrue, yPred):
2     y_shape = K.shape(yPred)
3     B = K.mean(K.square(yPred))
4     A = K.mean(yPred)
5     return -A/((B-A**2)**0.5)
```

Listing 8.26. Define negative Sharpe ratio as the customized loss function.

Given θ and $F_0 = 0$, at each time t, we evaluate the trading strategy F_t:

$$F_t = \text{sign}\left(\mu F_{t-1} + \sum_{i=0}^{p-1} \nu_i r_{t-i} + \omega\right),$$

where $F_t \in \{1, -1\}$, and $\theta := (\mu, \nu_0 \cdots, \nu_{p-1}, \omega)$ are model parameters.

```
1   def RRL_Model(input_dim, delta):
2     model = Sequential()
3     # initilize the input tensor with given shape (input_dim)
4     lagged_value = input_dim[-1]-1
5     T = input_dim[0]-1
6     input_layer = Input(shape= input_dim)
7     # Step 1: Split the input layer into two parts: X_t_layer and
      ↪   r_tplus1_layer
8     X_t_layer = Lambda(split_func, arguments={'p':lagged_value,
      ↪   'flag':1})(input_layer)
9     r_tplus1_Layer = Lambda(split_func, arguments={'p':lagged_value,
      ↪   'flag':2})(input_layer)
10    # Step 2: Map X_t_alyer to F_t_layer using SimpleRNN()
11    X_t_shape = X_t_layer.shape#[-2:]
12    F_t_Layer = SimpleRNN(1, input_shape= X_t_shape,activation = 'tanh',
      ↪   return_sequences = True, use_bias=True)(X_t_layer) #
13    # Step 3: Concatenate F_t_Layer with r_tplus1_Layer
14    hidden_layer = Concatenate()([F_t_Layer, r_tplus1_Layer])
15    output_layer = Lambda(trading_return, arguments={'delta':delta,
      ↪   'T':T})(hidden_layer )
16    model = Model(inputs=input_layer, outputs=output_layer)
17    sgd = optimizers.SGD(lr=0.01, decay=1e-6, momentum=0.9, nesterov=True)
18    model.compile(loss=sharpe_ratio_loss, optimizer=sgd)
19    return model
```

Listing 8.27. Python code for construction of the neural network for RRL.

As we cannot use the Keras function *predict*() to compute the next step trading strategy, we provide customized code for this in Listing 8.28.

8.4 Numerical Example: Algorithmic Trading

Following [Moody and Wu (1997)], let us consider the following synthetic dataset. We generate log price series as random walks with auto-regressive trend processes, i.e.,

$$p(t) = p(t-1) + \beta(t-1) + k\varepsilon(t),$$
$$\beta(t) = \alpha\beta(t-1) + \epsilon(t),$$

where α and k are constants, $(\varepsilon(t), \epsilon(t))$ is a two dimensional standard normal distributed variable. The artificial price series $z(t)$ is defined as

$$z(t) = \exp\left(\frac{p(t)}{R}\right),$$

```
1    from numpy import divide, power
2
3    def calc_next_window(weights, X_t, F_t):
4        """
5        Functionality: calculate the trading strategy at t+1.
6        Parameters:
7            weights: theta;
8            X_t: the feature at current time t
9            F_t: the trading strategy at current time t
10       Returns:
11           F_tplus1: the next trading strategy
12       """
13       weight0 = weights[0].reshape(np.shape(X_t))
14       F_tplus1 = np.sign(np.dot(weight0, X_t)+weights[1][0]*F_t + weights[2])
15       return F_tplus1
16
17   def do_prediction(X, model, delta):
18       weights = model.get_weights()
19       F = np.zeros(np.shape(X)[1], dtype = float)
20       R = np.zeros(np.shape(X)[1], dtype = float)
21       l = np.shape(F)[0]
22       for i in range(l):
23           if i == 0:
24               F[i] = calc_next_window(weights, X[0, i, :-1], 0)
25           else:
26               F[i] = calc_next_window(weights, X[0, i, :-1], F[i-1])
27               R[i] = F[i-1]*X[0, i-1, -1]-delta*(np.abs(F[i] - F[i-1]))
28       A = np.zeros(np.shape(F))
29       B = np.zeros(np.shape(F))
30       D = np.zeros(np.shape(F))
31       var_R = np.zeros(np.shape(F))
32       A[1:] = divide(np.cumsum(R)[1:], np.arange(l)[1:])
33       B[1:] = np.divide(np.cumsum(np.power(R, 2))[1:], np.arange(l)[1:])
34       var_R[2:] = B[2:]- power(A, 2)[2:]
35       D[2:] =  A[2:]/ np.sqrt(var_R[2:])
36       return F, R[:-1], D
```

Listing 8.28. Evaluation of RRL algorithm for incoming new data.

where $R := \max_t(p(t)) - \min_t(p(t))$ over a simulation with 10,000 samples, and R is a scale defined as the range of $p(t)$. In this example, we set the parameters $\alpha = 0.9$ and $k = 3$. The python code to simulate the artificial price based on the above model is given in Listing 8.29. A sample of the simulated price process is given in Figure 8.4.

```
1   import numpy as np
2   import matplotlib.pyplot as plt
3   np.random.seed(0)    # fix the random seed
4
5   def simulate_data(T):
6       b = np.zeros((T), float)
7       p = np.zeros((T), float)
8       b[0] = 0
9       p[0] = 0
10      for i in range(T-1):
11          p[i+1] = p[i] + b[i] + 3 * np.random.randn()
12          b[i+1] = 0.9 * b[i] + np.random.randn()
13      R = np.max(p) - np.min(p)
14      z = np.exp(p/R)
15      return z
16
17  T = 10000
18  z = simulate_data(T)
```

Listing 8.29. Simulate the artificial price process in [Moody and Wu (1997)].

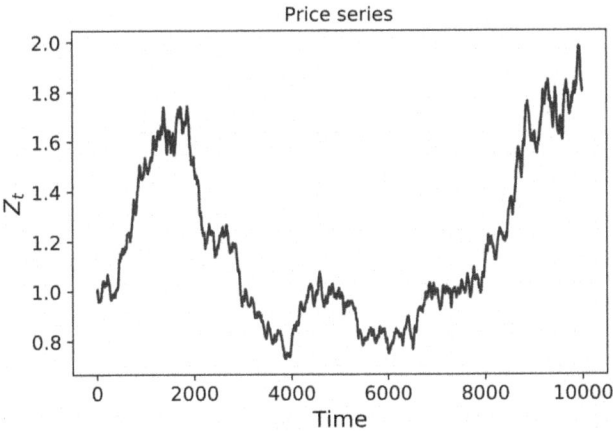

Figure 8.4. One sample trajectory of the artificial price process.

After simulation of the synthetic price data, the first step is to pre-process the data for our learning algorithm, which typically includes the following steps:

(1) Compute the return series $(r_t)_t$ and the p-lagged return series $(X_t)_t$ from the price data $(z_t)_t$ (Listing 8.30).
(2) Reshape and normalize the data $(X_t)_t$ (Listing 8.31).

```
1   import numpy as np
2   def construct_multi_feature_time_series(market_info, id_t_start, id_t_end,
    ↪  n_lagged_time_steps ):
3       """
4       Functionality:
5       To construct p-lagged return series X_ts and return series r_ts from
    ↪  price time series market_info from any time period with index ranging
    ↪  from id_t_start to id_t_end
6       """
7       # 1-dimensional price time series starting at id_t_start and ending at
    ↪  id_t_end
8       price = market_info[id_t_start:id_t_end]
9       returnIndex = np.diff(price) #simple return series
10      # Construct X_ts: the series of n_lagged_time_steps lagged values
11      X_ts = np.zeros((len(price) - n_lagged_time_steps - 1,
    ↪  n_lagged_time_steps))
12      for i in range(X_ts.shape[0]):
13          X_ts[i,:] = returnIndex[np.arange(i, i + n_lagged_time_steps,1)]
14      r_ts = returnIndex[n_lagged_time_steps:]
15      return X_ts, r_ts
16
17  X_ts, r_ts = construct_multi_feature_time_series(z, id_t_start=0,
    ↪  id_t_end=10000, n_lagged_time_steps=8)
```

Listing 8.30. Construct *p*-lagged return series and return series from price time series.

```
1   # create the feature set and the next return series from the price data
2   window_train = 2000
3   X_ts_train, r_ts_train = construct_multi_feature_time_series(z,
    ↪  id_t_start=0,  id_t_end = window_train, n_lagged_time_steps= 8)
4   X_ts, r_ts = construct_multi_feature_time_series(z, id_t_start= 0,
    ↪  id_t_end =10000, n_lagged_time_steps= 8)
5   # normalization
6   from sklearn import preprocessing
7
8   scaler = preprocessing.StandardScaler().fit(X_ts_train)
9   X_ts_train = scaler.transform(X_ts_train)
10  X_ts = scaler.transform(X_ts)
```

Listing 8.31. Compute the lagged values of return series and normalize the data.

The next step is to train the RRL model based on the training set (Listing 8.32) and then evaluate it on the testing set (Listing 8.33).

The numerical results are given in Figures 8.5 and 8.6. As shown in Figure 8.5, compared with buy-and-hold strategy, the RRL strategy

```
1   # Preparing the input data of the training set.
2   def reshape_input(X_ts, r_ts):
3       """ Parameters:
4       X_ts - lagged values of return series [T, p]
5       r_ts - the next value of return series [T, 1]
6           Return: [1, nTimeSteps, nFeatures], where nTimeSteps = T, nFeatures =
    ↪   p+1"""
7       X = np.concatenate([X_ts, r_ts], axis = 1)
8       X= np.reshape(X, (1,)+np.shape(X))
9       return X
10
11  # Preparing the input data of the training set
12  trainX = reshape_input(X_ts_train, r_ts_train)
13  X = reshape_input(X_ts, r_ts)
14
15  # Training an RRL model
16  delta = 0.002
17  model = RRL_Model(np.shape(trainX)[1:], delta)
18  print(model.summary())
19  model.fit(trainX[:window_train], trainX[:window_train], epochs=800,
    ↪   callbacks=callbacks, verbose=2)
```

Listing 8.32. Training process of RRL model.

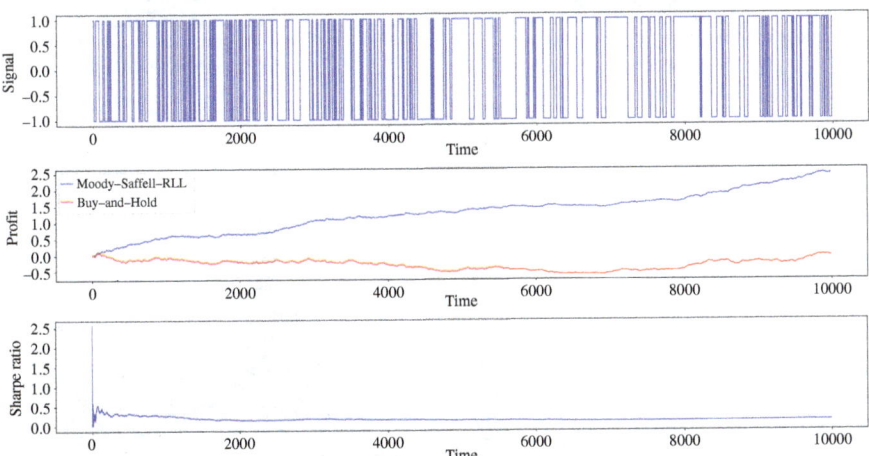

Figure 8.5. Trading signals (top panel), cumulative sums of profits (second panel) and moving average Sharpe ratio with $\eta = 0.01$ (bottom panel). The system performs relatively poorly while learning from scratch during the first 2000 time periods, but it improves thereafter.

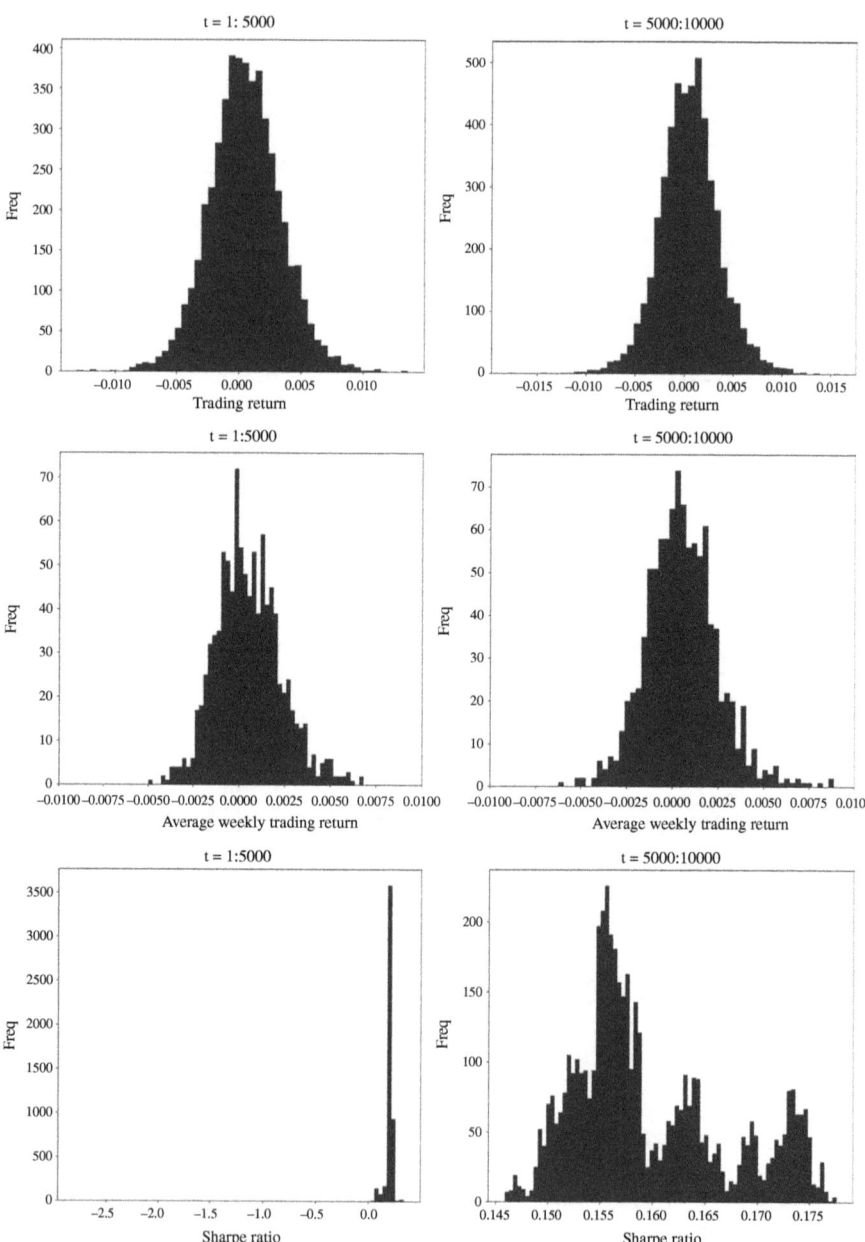

Figure 8.6. Histograms of the price changes (top), trading profits per time period (middle) and Sharpe ratios (bottom) for the simulation shown in Figure 8.5. The left column is for the first 5000 time periods and the right column is for the last 5000 time periods. The transient effects during the first 2000 time periods for real-time recurrent learning are evident in the lower left graph.

```
1  X = ReshapeInput(X_ts, r_ts)
2  F, R, D = Prediction(X, model, delta)
```

Listing 8.33. Evaluation of RRL algorithm on testing data.

produces higher total profit. Figure 8.6 demonstrates that RRL performs even better in the second half of the total trading period in terms of the daily return, the weekly profit and Shapiro ratio.

When we further investigate the effects of the transaction cost parameter δ in the RRL algorithm using the above code, we find that when δ is large, the above algorithm is not stable. One interesting line of inquiry is to use other loss function—e.g., the differential Sharpe Ratio—and see whether this further improves the efficiency of the algorithm. The relevant definitions are given in the exercises part.

In [Moody and Saffell (2001)], the application of RRL on the S&P 500 and T-bills is studied, which takes into account transaction fees and outperforms the Q-Trader voting strategy (using Q-learning) and a buy-and-hold strategy in the US market from 1970 to 1994. In [Deng *et al.* (2017)], a deep RNN for real-time financial signal representation is introduced to combine with RRL. The deep network plays an important role of automatically extracting features of the dynamic market conditions. Numerical results show that compared with RRL in [Moody and Saffell (2001)], deep reinforcement learning (DRL) makes much higher profits on the daily S&P index. It also raises the question about feature engineering, i.e., for empirical data, whether carefully chosen external variables improve the accuracy of DRL.

8.5   Exercises

(1) Following direct reinforcement learning [Moody and Saffell (2001)] in this chapter, write codes to generate 10,000 samples of price trajectories with a total of 1,000 time steps based on synthetic data. Implement the RRL algorithm for various transaction cost parameters δ, and show the box-plot of the P&L of the samples of the RRL algorithm in the testing dataset.

(2) Download the daily data of S&P from 1970 to 1994 and apply DRL to it. How can you compare the numerical results with other traditional trading strategies?

(3) Implement another type of reinforcement learning method to derive the optimal trading strategy (e.g., Q-learning).
(4) In [Moody and Saffell (2001)] the Exponential Moving Average Sharpe Ratio S_η and Differential Shapiro Ratio D_η are proposed to speed up the training of a RRL trading system. Implement customized loss functions for S_η and D_η in Keras. Apply them to the above synthetic data, and investigate the performance of RRL with the Sharpe ratio as the loss function.

Definition 8.1 (Exponential Moving Average Sharpe Ratio). An exponential moving average Sharpe ratio $S_\eta(t)$ on time scale η^{-1} is defined as follows:

$$S_\eta(t) = \frac{A_\eta(t)}{K_\eta(B_\eta(t) - A_\eta(t)^2)^{1/2}},$$

with $A_\eta(0) = B_\eta(0) = 0$, $A_\eta(t) = \eta R_t + (1 - \eta)A_\eta(t - 1)$ and $B_\eta(t) = \eta R_t^2 + (1 - \eta)B_\eta(t - 1)$.

Definition 8.2 (Differential Sharpe Ratio). A Differential Shapiro ratio $(D(t))$ on time scale η^{-1} is defined as follows:

$$D_\eta(t) \equiv \frac{B_\eta(t - 1)\Delta A_t - \frac{1}{2}A_\eta(t - 1)\Delta B_\eta(t)}{(B_\eta(t - 1) - A_\eta(t - 1)^2)^{3/2}},$$

with parameter η and A_η, B_η are defined in Definition 8.1.

Chapter 9

Case Study in Finance: Home Credit Default Risk

In this last chapter, we aim to equip readers with hands-on experience in the application of machine learning to real-world data challenges. The pipelines of machine learning applications include the main steps: (1) Pre-process the data; (2) Extract features; (3) Train a base model; (4) Parameter tuning; (5) Ensemble base models; and (6) Make final predictions (Figure 9.1).

This learning pipeline is an iterative procedure. We may start with one feature set and a base model to obtain its predictor. Based on the fitting results, we adjust the model architecture and feature sets to improve the base model and obtain various model candidates. Then we can select the

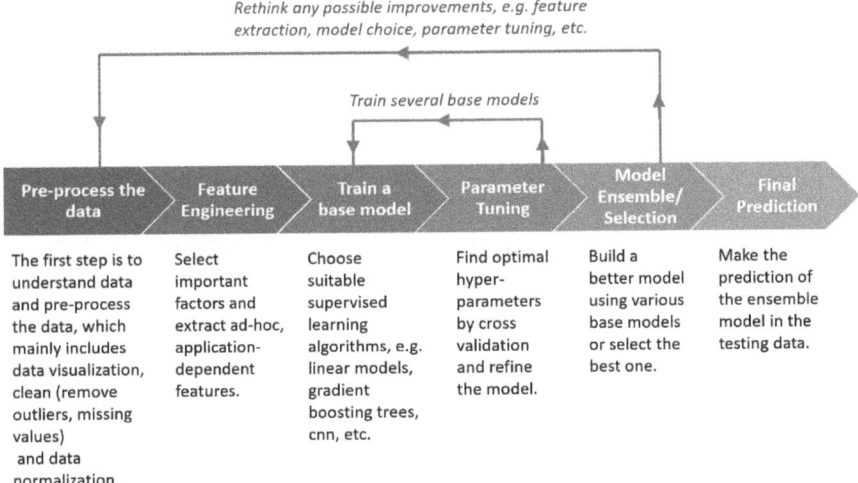

Figure 9.1. The pipelines of machine learning applications.

213

best model or employ model ensemble methods to construct a meta-model. Looking at the prediction results, we may rethink each step of the pipeline from pre-processing to model ensemble, and try every possible way to boost the performance. Lastly, we will use the optimal model(s) to make the final prediction in the testing set.

In the following, we will walk through the machine learning pipeline in one Kaggle competition on a financial application step by step as an example. We aim to help beginners in machine learning learn

- how to apply machine learning to solve empirical financial problems;
- the basics of how to get started in Kaggle challenges.

9.1   Problem Setup and Data

The chosen example is one of the most popular Kaggle competitions: Home Credit Default Risk,[1] which was recently hosted on Kaggle and has more than 7000 participants (individuals or teams) with a top prize of $35,000$.

The data for the Default Risk competition is provided by Home Credit,[2] which is a service provider offering credit (loans) to the unbanked population. Predicting whether or not a client will repay a loan is very critical to the business need. The main objective of this competition is to classify whether a client has the ability to repay their loan based on the history of their credit information. It is a binary classification task using multi-modal input data.

Let us first have a look at the data provided in this competition. A kernel entitled Start Here: A Gentle Introduction[3] by Will Koehrsen provides a nice description of the problem. It gives a summary of data structure shown in Figure 9.2. There are seven different sources of data:

- application_train/application_test.csv: the main training and testing data with information about each loan application. Every loan has its own row, with the unique identifier given by the feature 'SK_ID_CURR'. The training application data comes with 'TARGET' indicating 0: the loan was repaid or 1: the loan was not repaid. The testing application data do not include 'TARGET'.

---

[1]https://www.kaggle.com/c/home-credit-default-risk.
[2]http://www.homecredit.net/about-us.aspx.
[3]https://www.kaggle.com/willkoehrsen/start-here-a-gentle-introduction.

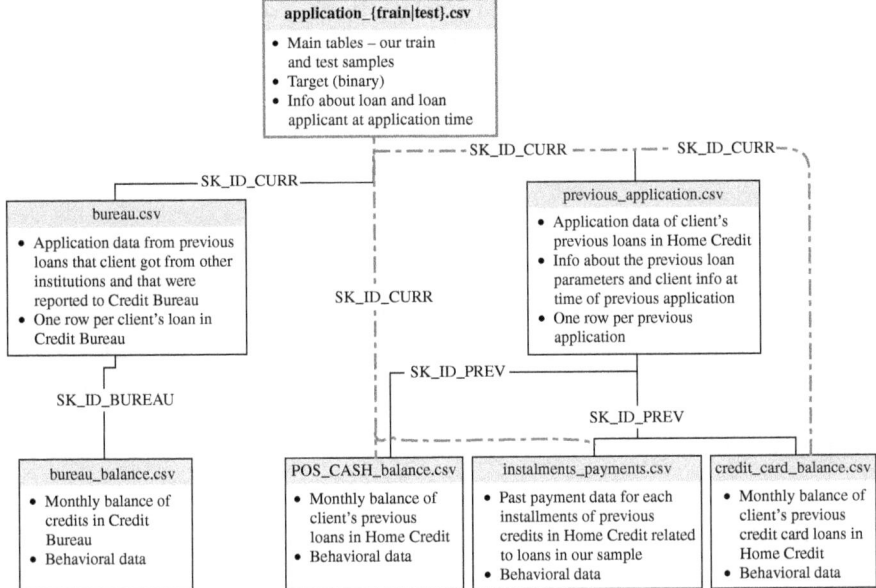

Figure 9.2. The structure of data provided in the Home Credit Default Risk competition.

- bureau.csv: data concerning clients' previous credit transactions reported to a Credit Bureau or obtained from other financial institutions.
- bureau_balance.csv: monthly data about the previous credits in bureau.csv.
- previous_application.csv: previous applications for loans at Home Credit of clients who have loans in the application data.
- POS_CASH_BALANCE.csv: monthly data about previous point of sale or cash loans clients have had with Home Credit.
- credit_card_balance.csv: monthly data about previous credit cards clients have had with Home Credit.
- installments_payment.csv: payment history for previous loans at Home Credit.

In the following, we only use application_train/application_test.csv, which contains most information among all the data.

In the Kaggle platform, for this competition, the participants must upload submission files in csv format. For each 'SK_ID_CURR' in the testing set, you must predict a probability for 'TARGET'. The file should contain a header and have the following format:

SK_ID_CURR	TARGET
100001	0.5
100005	0.5
100013	0.5
...	...

The performance measurement is AUC (see Section 2.2.4.3). The larger AUC is, the better the prediction. The ranking of participants is based on the AUC of the submissions, from highest to lowest. You can find this information in the evaluation section of this Kaggle competition.[4]

9.2 Exploratory Data Analysis

9.2.1 Imbalanced data

As this is a binary classification problem, let us first start with the percentage of class 0 and 1. The distribution of the target variable in the training set is given in Figure 9.3. Here you can see that the number of failed repayment applications is far less than that for successes.

9.2.2 Missing values

Next let us see how many missing data values there are in this example. Using the code provided in Figure 9.4, we see that the top five factors that have most missing data include 'COMMONAREA_MODE', 'COMMONAREA_MEDI' and 'COMMONAREA_AVG'.

9.2.3 Feature grouping

In Application.csv data, there are 122 input factors. For convenience, we group them based on the contents of the input factors. The data is grouped into the following main groups (this is not an exhaustive list of groups. Interested readers should refer to the github code):

---

[4]https://www.kaggle.com/c/home-credit-default-risk/overview/evaluation.

Code:

```
1  import seaborn as sns
2  df_train = df_all[df_all['TARGET'].notnull()]
3  df_train['TARGET']=df_train['TARGET'].astype(int) ax =
4  sns.countplot(df_train['TARGET'])
```

Output:

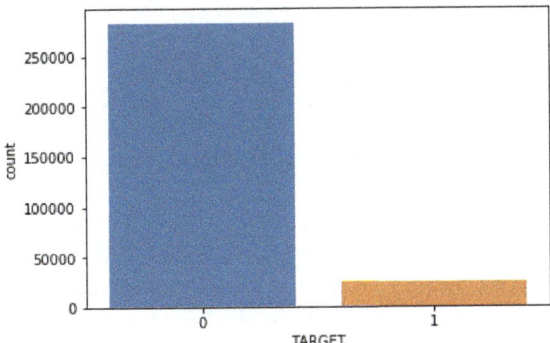

Figure 9.3. Imbalanced data.

Code:

```
1  def missing_data(data):
2      total = data.isnull().sum().sort_values(ascending = False)
3      percent = data.isnull().sum()/data.isnull().count()*100
4      percent = percent.sort_values(ascending = False)
5      return pd.concat([total, percent], axis=1, keys=['Total', 'Percent'])
6
7  print(missing_data(df_train))
```

Output:

Feature	Total	Percent
COMMONAREA_MODE	214865	69.87%
COMMONAREA_MEDI	214865	69.87%
COMMONAREA_AVG	214865	69.87%
NONLIVINGAPARTMENTS_MODE	213514	69.43%
NONLIVINGAPARTMENTS_AVG	213514	69.43%

Figure 9.4. The code for computing the percentage of missing data and the snapshot of the output result. The top five factors with highest percentage of missing data.

- External source features (sel_feas_EXT_SOURCE):
  - 'EXT_SOURCE_1';
  - 'EXT_SOURCE_2';
  - 'EXT_SOURCE_3'.
- AMT related features (sel_feas_AMT):
  - 'AMT_ANNUITY';
  - 'AMT_CREDIT';
  - 'AMT_INCOME_TOTAL';
  - 'AMT_GOODS_PRICE'.
- Personal information (sel_feas_PERSON): e.g., gender, contract type, house type, etc.

In this dataset, we have 65 float input factors, 42 integer input factors and 16 object types factors. We have multi-modal data, and the learning algorithm we choose should be able to deal with both numerical variables and categorical variables.

9.3   Building the First Classifier

In this section, we implement LightGBM for this binary classfication using only the application data. LightGBM is a very popular gradient boosting tree model, which has the advantage of high efficiency compared with other gradient boosting methods.

9.3.1   Data preprocessing

Preprocessing the dataset usually includes the following steps:

(1) Convert the data into the types that the learning algorithm can process;
(2) Split the data into the training and testing data;
(3) Handle the missing data;
(4) Remove outliers;
(5) Normalize the data.

The LightGBM algorithm, which we are using in this example, can be directly used with missing data and is not sensitive to normalization. Therefore the preprocessing here mainly requires splitting the training and testing data.

9.3.2 Feature engineering

In practice, it is very important to find informative features to feed into the learning algorithm, which is called feature engineering. In some problems, the dimensionality of the raw data is very high relative to the sample size, which can easily cause severe overfitting issues. For any data problem, we should carefully select an appropriate subset of the input or find informative transformations of the raw data into features, which requires domain knowledge. Failures in feature engineering may lead to very poor predictive results.

In this example, we have two main approaches for the customized feature set:

- features by aggregation;
- features by domain knowledge.

The implementation of feature calculation is given in Listing 9.34. Now we give you some examples of each type of feature. For an aggregated feature, instead of considering every factor associated with 'ADDRESS_MATCH', we use the sum of the factors as a single valued feature called 'ADDRESS_MATCH_SCORE_TOTAL'. For a domain knowledge derived feature, a good example is the annuity percentage, which is defined as the ratio of the annuity of the applicant to his/her credit. It better reflects how likely the applicant is to repay the loan than simply the level of annuity or credit.

After feature engineering, we have 40 features for this binary classification problem, which is significant dimension reduction (from 122 input factors) and is expected to preserve the main information of the raw input data based on our domain knowledge.

9.3.3 Training a model

LightGBM is a gradient boosting method that uses tree-based models. The main difference between LightGBM and other tree-based methods is the growth direction of the trees. LightGBM grows the tree leaf-wise while other algorithms grow level-wise (see Figure 9.5). At each iteration, LightGBM splits a leaf so as to decrease the loss function most.

```
1   def features_add_derivative(df_all, feature_groups):
2       '''
3       derivativeI: features from aggregation
4       derivativeII: features from domain knowledge
5       '''
6       # aggregation
7       df = df_all.copy()
8       agg_mapping = {
9           'DOCUMENT': 'FLAG_DOCUMENT_TOTAL',
10          'AMT_REQ': 'AMT_REQ_CREDIT_BUREAU_TOTAL',
11          'BUILDING': 'BUILDING_SCORE_TOTAL',
12          'CONTACT': 'CONTACT_SCORE_TOTAL',
13          'ADDRESS_MATCH': 'ADDRESS_MATCH_SCORE_TOTAL'}
14      agg_func = lambda group_name, df:
        ↪ df[feature_groups[group_name]].sum(skipna=True, axis=1)
15      for (group_name, new_name) in agg_mapping.items():
16          df[new_name] = agg_func(group_name, df)
17      # new features: from domain knowledge
18      df['AMT_INCOME_TOTAL'].replace(1.170000e+08, np.nan, inplace=True)
19      df['ANNUITY_CREDIT_PERC'] = df['AMT_ANNUITY'] / df['AMT_CREDIT']
20      df['ANNUITY_INCOME_PERC'] = df['AMT_ANNUITY'] / df['AMT_INCOME_TOTAL']
21      df['GOODS_PRICE_CREDIT_PERC'] = df['AMT_GOODS_PRICE'] /
        ↪ df['AMT_CREDIT']
22      sel_feas_object = [col for col in df.columns if df[col].dtype ==
        ↪ 'object']
23      # Categorical features: Binary features and One-Hot encoding
24      for bin_feature in (sel_feas_object):
25          df[bin_feature], uniques = pd.factorize(df[bin_feature])
26      return df
```

Listing 9.34. Code for deriving the new features by aggregation and domain knowledge.

There is also a LightGBM Python package developed by Microsoft for fitting and prediction using the LightGBM model.[5] In Listing 9.35, we construct a LightGBM model using the LightGBM package.

9.3.4 Out-of-folds prediction

In this subsection, let us show how to implement OOF prediction as discussed in Section 2.3. We define a customized function called train_model_lgbm(), which takes the training set and testing set, the id number vector, the folds, the parameter set of the LightGBM classifier and the parameter set of the fitting as the function input. The return values

---

[5]https://lightgbm.readthedocs.io/en/latest/index.html#.

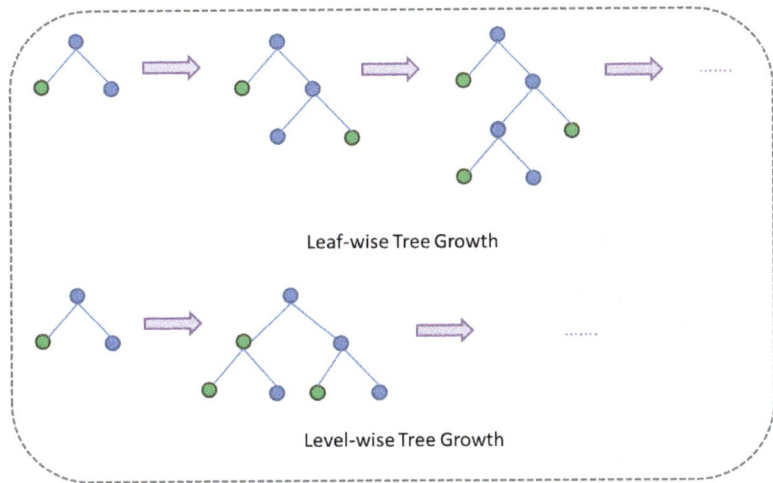

Figure 9.5. The difference between LightGBM and other tree-based methods.

of train_model_lgbm() are a collection of estimated models oof_models (see Listing 9.36).

In Listing 9.37, we use stratified K-fold cross validation to implement OOF prediction. Table 9.1 provides the AUC score of each fold, which is consistently around 77%. The ROC curve of the testing set is given in Figure 9.6, and the average ROC AUC is 0.7699 with standard deviation 0.0050.

The average feature importance of OOF is provided in Figure 9.7. Figure 9.7 indicates that the most important factors are ANNUITY_CREDIT_PERC, EXT_SOURCE_3 and DAYS_CREDIT_PERC. This fits the intuition that those factors have strong predictive power for forecasting the repayment ability of the applicants. Note that ANNUITY_CREDIT_PERC is the feature designed by domain knowledge, which indicates the importance of feature engineering. Besides this, the EXT_SOURCE features are strong factors.

9.3.5 Parameter tuning

We apply cross-validation and grid search to choose the optimal hyper-parameters in Listing 9.38. As the scikit-learn built-in function gridsearch_cv() does not support early stopping, we define a customized grid search function shown in Listing 9.38. Then we apply lbg.cv to compute the results of the cross-validation for parameter tuning.

```
1   import lightgbm as lgb
2   _default_algo_params_lgbm = {'objective': 'binary',
3               'metric': 'auc',
4               'num_threads': 6,
5               'num_iterations': 10000,
6               'max_depth': 5,
7               'learning_rate': 0.03,
8               'bagging_fraction': 0.744,
9               'feature_fraction': 0.268,
10              'lambda_l1': 0.91,
11              'lambda_l2': 0.89,
12              'min_child_weight': 18.288,
13              'min_gain_to_split': 0.0365,
14              'verbose': -1,
15              'silent': -1}
16
17  _default_fit_params_lgbm = {
18      "eval_metric": 'auc',
19      'verbose': 1000,
20      'early_stopping_rounds': 100
21  }
22  def Lgbm_fit(trn_x, tran_y, val_x, val_y, algo_parames, fit_params):
23      clf = lgb.LGBMClassifier(**algo_params)
24      fit_params.update({"eval_set": [(trn_x, trn_y), (val_x, val_y)]})
25      clf.fit(trn_x, trn_y, **fit_params)
26      return clf
27
28  clf = Lgbm_fit(trainX, tranY, valX, valY, _default_algo_params_lgbm,
    ↪  _default_fit_params_lgbm)
```

Listing 9.35. Constructing a LightGBM model.

According to Table 9.2, the best hyper-parameter set is with a learning rate of 0.03, max depth of tree of 6 and number of leaves of 31, which has the highest five-fold average accuracy of 0.770211. Interested readers may refer to the official documentation for more information about parameter tuning in the LightGBM model.[6]

9.4 Model Stacking

To obtain an accurate and robust estimator for the testing dataset, here we first consider using three different models, i.e., LightGBM, logistic and

---

[6]https://lightgbm.readthedocs.io/en/latest/Parameters-Tuning.html.

```
1  def train_model_lgbm(data_, test_, y_, ids, folds_, algo_params,
   ↪  fit_params):
2      oof_models = []
3      feats = [f for f in data_.columns if f not in ['SK_ID_CURR']]
4      for n_fold, (trn_idx, val_idx) in enumerate(folds_.split(data_, y_)):
5          trn_x, trn_y = data_[feats].iloc[trn_idx], y_.iloc[trn_idx]
6          val_x, val_y = data_[feats].iloc[val_idx], y_.iloc[val_idx]
7          clf = lgb.LGBMClassifier(**algo_params)
8          fit_params.update({"eval_set": [(trn_x, trn_y), (val_x, val_y)]})
9          clf.fit(trn_x, trn_y, **fit_params)
10         oof_models.append(copy.deepcopy(clf))
11         del clf, trn_x, trn_y, val_x, val_y
12         gc.collect()
13     folds_idx = [(trn_idx, val_idx) for trn_idx, val_idx in
   ↪      folds_.split(data_, y_)]
14     res_models = {'oof_models': oof_models, 'folds_idx':folds_idx}
15     return res_models
```

Listing 9.36. Training the LightGBM model on the training set.

```
1  def oof_prediction_test(data_, test_, y_, ids, folds_, res_models):
2      '''
3      functionality:
4      oof_prediction + predicatin on the test set.
5      '''
6      sub_preds = np.zeros(test_.shape[0]) # test prediction
7      feats = [f for f in data_.columns if f not in ['SK_ID_CURR']]
8      for n_fold, (trn_idx, val_idx) in enumerate(folds_.split(data_, y_)):
9          clf = res_models['oof_models'][n_fold]
10         sub_preds += clf.predict_proba(
11             test_[feats],
12             num_iteration=clf.best_iteration_)[:, 1] / folds_.n_splits
13         test_['TARGET'] = sub_preds
14         del clf
15         gc.collect()
16     test_['TARGET'] = sub_preds
17     folds_idx = res_models['folds_idx']
18     res = {
19         'y': y_,
20         'folds_idx': folds_idx,
21         'test_preds': test_[['SK_ID_CURR', 'TARGET']],
22     }
23     return res
```

Listing 9.37. OOF prediction using stratified *K*-fold cross validation.

Table 9.1. OOF prediction.

	Fold 1	Fold 2	Fold 3	Fold 4	Fold 5	Full
AUC	0.7684	0.7701	0.7738	0.7614	0.7758	0.7699

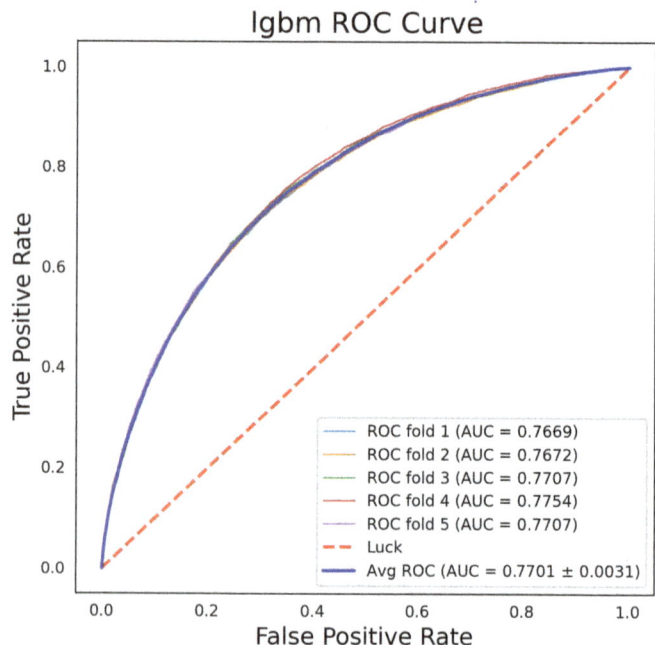

Figure 9.6. ROC curve for five-fold cross validation.

neural network as the base model. Then we use stacking, discussed in Section 2.3.3, for model ensemble. A Kaggle kernel entitled *A Kaggler's Guide to Model Stacking in Practice*[7] provides a nice tutorial for model stacking. Interested readers may refer to it for details.

In Section 9.3.5, we showed how to train a LightGBM model and implement the prediction using OOF prediction. For the logistic model and neural network, we can perform a similar implementation, simply modifying the model and training. This motivates us to implement the code in a generic way to avoid repetition of the common parts of these three models.

---

[7]https://datasciblog.github.io/2016/12/27/a-kagglers-guide-to-model-stacking-in-practice/.

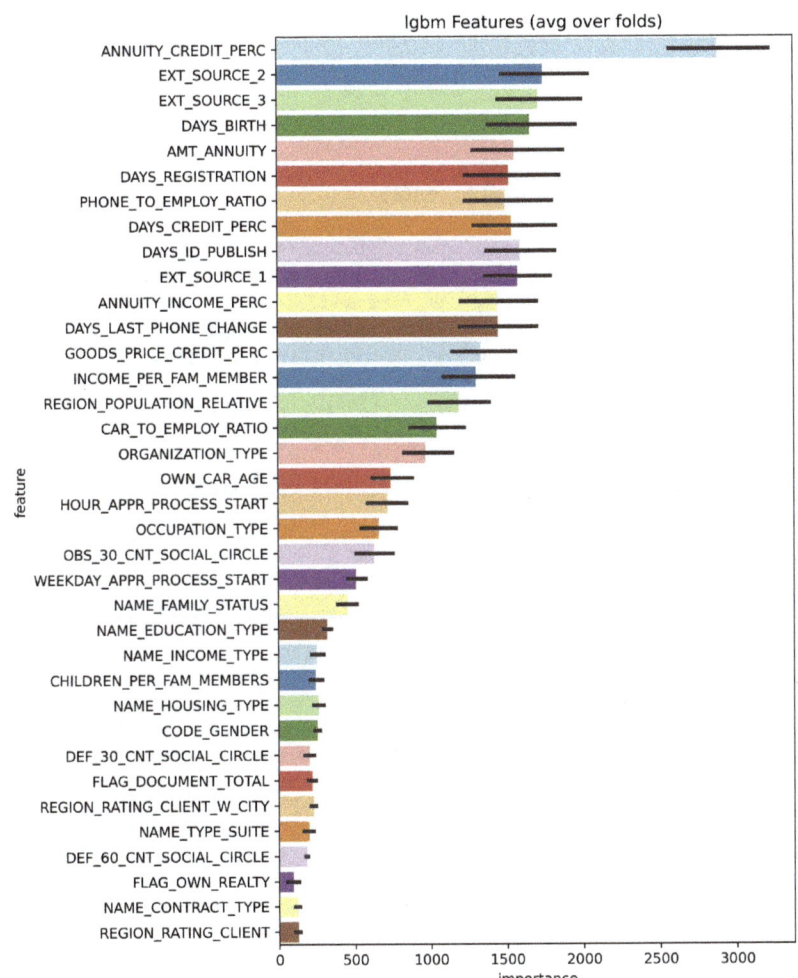

Figure 9.7. Feature importance.

Figure 9.8 shows the prediction of the testing data using LightGBM with two different parameters (denoted by **lgb1** and **lgb2**), a logistic model (denoted by **lm**) and a neural network model (denoted by **nn**).

The Python project for the credit default forecast application is constructed as follows:

- preprocessing.py:
  - load input data;
  - feature selection.

```
1  def GridSearch(x_train, sel_feas, y_train, param_grid, algo_params =
   ↪  _default_algo_params_lgbm, cv = 5, num_boost_round=  2000,
   ↪  early_stopping_rounds=100, verbose_eval=1000, seed = 5):
2      best_params = {}
3      grid_scores = []
4      for i in range(len(param_grid)):
5          algo_params.update(param_grid[i])
6          # cross valition for one set of parameters
7          cvresult = lgb.cv(algo_params, lgb.Dataset(x_train[sel_feas],
   ↪      label=y_train.values.astype(int)), nfold=cv, stratified=True,
   ↪      num_boost_round= num_boost_round,
   ↪      early_stopping_rounds=early_stopping_rounds,
   ↪      verbose_eval=verbose_eval, seed = seed, show_stdv=True)
8          # k-fold cross validation accuracy and std
9          curr_auc = pd.Series(cvresult['auc-mean']).max()
10         curr_std = pd.Series(cvresult['auc-mean']).std()
11         cv_res = param_grid[i]
12         cv_res.update({'auc': curr_auc, 'std': curr_std})
   ↪      grid_scores.append(cv_res)
13         if curr_auc > max_auc:
14             best_params = param_grid[i]
15     return best_params,  grid_scores
```

Listing 9.38. Grid search for parameter tuning.

Table 9.2. Numerical results from grid Search and cross validation.

learning rate = 0.01		
Max depth / No of leaves	5	6
20	0.769454	0.769392
31	0.769208	0.769523
learning rate = 0.03		
Max depth / No of leaves	5	6
20	0.769797	0.769792
31	0.769995	**0.770211**

- model_train.py:

    – train model: lgbm, logistic, neural network; train results (pass KFold to model); save results.

	SK_ID_CURR	lgb1	lgb2	Lm	nn
0	100001	0.029291	0.037438	0.082431	0.044820
1	100005	0.090089	0.086825	0.103101	0.156741
2	100013	0.012217	0.015666	0.086690	0.021608
3	100028	0.049396	0.045507	0.094616	0.029320
4	100038	0.127582	0.129161	0.108598	0.129872

Figure 9.8. The data frame constructed by four predictors using different models.

Table 9.3. Summary of the AUC score of base models and meta-model1.

	lgb1	lgb2	lm	nn	Model Stacking 1
Test AUC	0.77047	0.76476	0.66799	0.74297	0.77085

- model_helper.py:
  - feature importance;
  - plot: importance, roc_curve, precision.
- several main files:
  - main_run_base_model.py: run four base models;
  - main_grid_search.py: grid search for parameter tuning;
  - main_model_stack.py: model stacking.
- run_model_kaggle_home_credit.ipynb:
  - construct default model parameters;
  - example of running model training and save result.

We use LightGBM to construct a meta-model for the functional relationship between the four predictors and the actual output label. Table 9.3 gives the comparison results of the AUC scores of the base models and the meta-model using stacking. After model stacking, the meta-model is slightly better than that of all base models, but the improvement may not be statistically significant. One possible reason for this is that stacking performs very well when base models have less dependency. However, in our case, the features of the data for the four base models are the same, which gives the four predictors strong co-dependency.

Table 9.4 gives the feature importance of each base model, and it suggests that the meta-model learns to put more weight on the better base models (i.e., **lgb1** and **lgb2**).

Table 9.4. Feature importance of meta-model.

	lgb1	lgb2	lm	nn
Feature importance	173.8	90.82	80.43	75.0

Table 9.5. Summary of the AUC scores of signature based models and meta-model2.

	sig1	sig2	Model Stacking 2
Test AUC	0.63228	0.58226	0.77833

To further boost the learning performance, we use installments_payments.csv and POS_CASH_balance.csv data, and an innovative feature set—called the signature feature set' (this is out of the scope of this book).[8] Table 9.5 shows that using the signature feature of the installments_payments.csv and POS_CASH_balance.csv with lightGBM separately (denoted by **sig1** and **sig2**), we obtained about 60% AUC in the testing data, which are relatively weak estimators compared with that of **lgb1** and **lgb2**. But it improves the AUC in the testing data from 0.77085 to 0.77833 by combining these with the previous base models by model stacking (denoted by **Model Stacking 2**). This demonstrates that that more diversity the base model has, the better results model stacking brings, even though some of the base models are weak estimators.

9.5   Result Submission to Kaggle

Let us see how well our model can perform on the testing data. We need to submit it to the Kaggle platform. In this case, we get a score of 0.77085. Bingo! It ranks in 4472nd position among all 7198 participants. You may think that the rank is not that high, but remember that we only used the application data. Now it is your turn to try your own method, and let us see how well you can do!

---

[8]Interested readers may refer to [Levin *et al.* (2013)]. The esig Python package provides a convenient toolset to compute the signature features of time series (https://esig.readthedocs.io/en/latest/).

9.6 Exercises

9.6.1 CFM challenge: Volatility forecast

The goal of the challenge is to predict the end-of-day volatility for American stocks knowing their past price history's features. This data challenge is sponsored by CFM, a leading global quantitative and systematic asset management firm. Founded in 1991, CFM is specialized in developing trading strategies based on a global and quantitative approach to financial markets. CFM's methodology relies on in-depth statistical analysis and the modelization of terabytes of financial data for asset allocation, trading decisions and order execution.

Challengers are provided with trading data between 9:30 am and 2:00 pm and asked to predict the volatility between 2:00 pm and 4:00 pm. The data provided contains two training files and one testing file.

The metric used in this challenge to designate the winning participant is Mean Absolute Percentage Error (MAPE). The ranking of the submitted results is determined by the MAPE for the testing data.

9.6.2 Other Kaggle competitions on financial applications

We list two popular past Kaggle competitions on financial applications, which are left to readers as exercises.

(1) *Two Sigma: Using News to Predict Stock Movements—Use news analytics to predict stock price performance.*[9] This Kaggle Challenge aims to address the question of whether we can use the content of news analytics to predict stock price performance.
(2) *Prudential: In a one-click shopping world with on-demand everything, the life insurance application process is antiquated. The main objective of this competition is to build an accurate predictive model for risk classification using extensive customer information.*[10]

---

[9]https://www.kaggle.com/c/two-sigma-financial-news.
[10]https://www.kaggle.com/c/prudential-life-insurance-assessment.

Bibliography

Ache, A. G. and Warren, M. W. (2019). Ricci curvature and the manifold learning problem, *Advances in Mathematics* **342**, pp. 14–66.

Baldi, P. and Sadowski, P. J. (2013). Understanding dropout, in *Advances in Neural Information Processing Systems*, pp. 2814–2822.

Bhandare, A., Bhide, M., Gokhale, P., and Chandavarkar, R. (2016). Applications of convolutional neural networks, *International Journal of Computer Science and Information Technologies* **7**, 5, pp. 2206–2215.

Breiman, L. (1996). Bagging predictors, *Machine Learning* **24**, 2, pp. 123–140.

Breiman, L. (1999). Pasting small votes for classification in large databases and on-line, *Machine Learning* **36**, 1–2, pp. 85–103.

Chipman, H. A., George, E. I., and McCulloch, R. E. (1998). Bayesian cart model search, *Journal of the American Statistical Association* **93**, 443, pp. 935–948.

Cho, K., Van Merriënboer, B., Gulcehre, C., Bahdanau, D., Bougares, F., Schwenk, H., and Bengio, Y. (2014). Learning phrase representations using rnn encoder-decoder for statistical machine translation, *arXiv preprint arXiv:1406.1078*.

Deng, Y., Bao, F., Kong, Y., Ren, Z., and Dai, Q. (2017). Deep direct reinforcement learning for financial signal representation and trading, *IEEE Transactions on Neural Networks and Learning Systems* **28**, 3, pp. 653–664.

Ester, M., Kriegel, H.-P., Sander, J., Xu, X., *et al.* (1996). A density-based algorithm for discovering clusters in large spatial databases with noise, in *KDD-96 Proceedings*, Vol. 96, pp. 226–231.

Friedman, J., Hastie, T., and Tibshirani, R. (2001). *The Elements of Statistical Learning*, Vol. 1 (Springer series in statistics New York).

Friedman, J. H. (2001). Greedy function approximation: A gradient boosting machine, *Annals of Statistics*, pp. 1189–1232.

Goodfellow, I., Pouget-Abadie, J., Mirza, M., Xu, B., Warde-Farley, D., Ozair, S., Courville, A., and Bengio, Y. (2014). Generative adversarial nets, in *Advances in Neural Information Processing Systems*, pp. 2672–2680.

Graham, B. (2014). Fractional max-pooling, *arXiv preprint arXiv:1412.6071*.

Graves, A., Mohamed, A.-r., and Hinton, G. (2013). Speech recognition with deep recurrent neural networks, in *2013 IEEE International Conference on Acoustics, Speech and Signal Processing (IEEE)*, pp. 6645–6649.

Gunning, D. (2017). Explainable Artificial Intelligence (XAI). Defense Advanced Research Projects Agency (DARPA). Accessed on 2017-09-09. `http://www.darpa.mil/program/explainable-artificial-intelligence`.

Gybenko, G. (1989). Approximation by superposition of sigmoidal functions, *Mathematics of Control, Signals and Systems* **2**, 4, pp. 303–314.

Hastie, T., Rosset, S., Zhu, J., and Zou, H. (2009). Multi-class adaboost, *Statistics and its Interface* **2**, 3, pp. 349–360.

He, K., Zhang, X., Ren, S., and Sun, J. (2016). Deep residual learning for image recognition, in *Proceedings of the IEEE Conference on Computer Vision and Pattern Recognition*, pp. 770–778.

Hochreiter, S. and Schmidhuber, J. (1997). Long short-term memory, *Neural Computation* **9**, 8, pp. 1735–1780.

Hoff, P. D. (2017). Lasso, fractional norm and structured sparse estimation using a hadamard product parameterization, *Computational Statistics & Data Analysis* **115**, pp. 186–198.

Huang, G., Liu, Z., Van Der Maaten, L., and Weinberger, K. Q. (2017). Densely connected convolutional networks, in *Proceedings of the IEEE Conference on Computer Vision and Pattern Recognition*, pp. 4700–4708.

Huang, R. and Polak, T. (2011). Lobster: Limit order book reconstruction system, *Available at SSRN 1977207*.

Hubel, D. H. and Wiesel, T. N. (1962). Receptive fields, binocular interaction and functional architecture in the cat's visual cortex, *The Journal of Physiology* **160**, 1, pp. 106–154.

Ke, G., Meng, Q., Finley, T., Wang, T., Chen, W., Ma, W., Ye, Q., and Liu, T.-Y. (2017). Lightgbm: A highly efficient gradient boosting decision tree, in *Advances in Neural Information Processing Systems*, pp. 3146–3154.

Krizhevsky, A., Sutskever, I., and Hinton, G. E. (2012). Imagenet classification with deep convolutional neural networks, in *Advances in Neural Information Processing Systems*, pp. 1097–1105.

LeCun, Y., Bengio, Y., and Hinton, G. (2015a). Deep learning, *Nature* **521**, 7553, p. 436.

LeCun, Y. *et al.* (2015b). Lenet-5, convolutional neural networks, Available at: http://yann.lecun.com/exdb/lenet, p. 20.

Levin, D., Lyons, T., and Ni, H. (2013). Learning from the past, predicting the statistics for the future, learning an evolving system, *arXiv preprint arXiv:1309.0260*.

Lord, R. and Pelsser, A. (2007). Level–slope–curvature—fact or artifact? *Applied Mathematical Finance* **14**, 2, pp. 105–130.

Maks, A. *et al.* (1972). *An Introduction to Linear Algebra and Tensors* (Courier Corporation).

McCulloch, W. S. and Pitts, W. (1943). A logical calculus of the ideas imma-
nent in nervous activity, *The Bulletin of Mathematical Biophysics* **5**, 4,
pp. 115–133.

Mnih, V., Kavukcuoglu, K., Silver, D., Graves, A., Antonoglou, I., Wierstra, D.,
and Riedmiller, M. (2013). Playing atari with deep reinforcement learning,
arXiv preprint arXiv:1312.5602.

Moody, J. and Saffell, M. (2001). Learning to trade via direct reinforcement, *IEEE
Transactions on Neural Networks* **12**, 4, pp. 875–889.

Moody, J. and Wu, L. (1997). Optimization of trading systems and portfolios,
in *Computational Intelligence for Financial Engineering (CIFEr), 1997.,
Proceedings of the IEEE/IAFE 1997 (IEEE)*, pp. 300–307.

Rosenblatt, F. (1957). *The Perceptron, a Perceiving and Recognizing Automaton
Project Para* (Cornell Aeronautical Laboratory).

Silver, D., Huang, A., Maddison, C. J., Guez, A., Sifre, L., Van Den Driessche, G.,
Schrittwieser, J., Antonoglou, I., Panneershelvam, V., Lanctot, M., *et al.*
(2016). Mastering the game of go with deep neural networks and tree search,
Nature **529**, 7587, pp. 484–489.

Simonyan, K. and Zisserman, A. (2014). Very deep convolutional networks for
large-scale image recognition, *arXiv preprint arXiv:1409.1556*.

Sirignano, J. and Cont, R. (2018). Universal features of price formation in financial
markets: Perspectives from deep learning.

Srivastava, N., Hinton, G., Krizhevsky, A., Sutskever, I., and Salakhutdinov, R.
(2014). Dropout: a simple way to prevent neural networks from overfitting,
The Journal of Machine Learning Research **15**, 1, pp. 1929–1958.

Steinkraus, D., Buck, I., and Simard, P. (2005). Using gpus for machine learning
algorithms, in *Eighth International Conference on Document Analysis and
Recognition (ICDAR'05) (IEEE)*, pp. 1115–1120.

Sutton, R. S. and Barto, A. G. (2018). *Reinforcement Learning: An Introduction*
(MIT press).

Szegedy, C., Liu, W., Jia, Y., Sermanet, P., Reed, S., Anguelov, D., Erhan, D.,
Vanhoucke, V., and Rabinovich, A. (2015). Going deeper with convolutions,
in *Proceedings of the IEEE Conference on Computer Vision and Pattern
Recognition*, pp. 1–9.

Thorndike, R. L. (1953). Who belongs in the family? *Psychometrika* **18**, 4,
pp. 267–276.

Thrun, S. and Pratt, L. (2012). *Learning to Learn* (Springer Science & Business
Media).

Tibshirani, R., Walther, G., and Hastie, T. (2001). Estimating the number of
clusters in a data set via the gap statistic, *Journal of the Royal Statistical
Society: Series B (Statistical Methodology)* **63**, 2, pp. 411–423.

Van Den Oord, A., Dieleman, S., Zen, H., Simonyan, K., Vinyals, O., Graves, A.,
Kalchbrenner, N., Senior, A. W., and Kavukcuoglu, K. (2016). Wavenet:
A generative model for raw audio. *SSW* **125**.

Wang, S. and Jiang, J. (2015). Learning natural language inference with lstm, *arXiv preprint arXiv:1512.08849*.

Ward Jr, J. H. (1963). Hierarchical grouping to optimize an objective function, *Journal of the American Statistical Association* **58**, 301, pp. 236–244.

Zhang, Q.-s. and Zhu, S.-C. (2018). Visual interpretability for deep learning: a survey, *Frontiers of Information Technology & Electronic Engineering* **19**, 1, pp. 27–39.

Index

A

accuracy, 36–37, 40, 43, 45, 123, 128, 146, 150, 152

activation function, 89–90, 92, 99, 120

adjusted R^2, 31, 53

alternating ridge regressions, 63

artificial neural network (ANN), 9–10, 94, 98–101, 105, 107–110, 118, 129–130, 133

AUC, 41, 227

average silhouette width, 160

B

backpropagation, 87, 100

backpropagation algorithm, 104

backpropagation through time (BPTT), 133, 137

bagging, 45, 50, 78

basis expansion, 9, 17, 64

batch GD, 29

batch gradient descent (BGD), 20, 24–25, 28

between-cluster variance, 158

big data, 1, 5

binary classification, 37, 39, 144, 216

binary tree, 71, 73

black box learning, 7

boosting, 44, 46–47, 50

C

categorical variable, 33–34

Cifar10, 118, 122–123, 126

classification, 15, 33, 34, 120

classification and regression tree (CART), 72

classification tree, 74

cluster analysis, 9

clustering and PCA, 194

clustering methods, 154, 156, 166, 169

common factors, 184, 186

complete linkage (CL), 164

confusion matrices, 83

confusion matrix, 37, 40, 85, 108

connectivity, 89, 117, 122, 166–167, 169

convolution operator, 117

convolutional layer, 89, 112–113, 116–117, 120, 122

convolutional neural network (CNN), 9–10, 109–110, 112, 117–118, 121–122, 125, 128–129

convolutional operator, 115–116

cost function, 19

cross entropy, 35, 74, 123

cross-validation, 32–33, 130, 221, 223

curvature change, 185

D

data augmentation, 124, 127–128

data fusion, 8

data preprocessing, 218

data segmentation, 194–195

data streams, 2

data-driven, 3

decision tree, 9, 68, 73–74, 80–81, 84–86

deep learning, 8, 87, 89, 110

deep neural networks, 91

dendrogram, 163, 165, 174

dense layer, 118, 120

density-based clustering, 154, 157

density-based spatial clustering of applications with noise (DBSCAN), 154, 166–169